FLOWERING PLANTS OF THAILAND
A FIELD GUIDE

FLOWERING PLANTS OF THAILAND
A FIELD GUIDE

Patrick D. McMakin

White Lotus
Bangkok   Cheney

Publisher's Note

The text of this book is identical to that of *A Field Guide to the Flowering Plants of Thailand* but it has been completely re-typeset in a more compact format.

© 1993 by Patrick D. McMakin, all rights reserved
First edition 1988
Second edition 1993

White Lotus Co.,Ltd.
P.O. Box 1141
Bangkok, Thailand

Printed in Thailand

Typeset by COMSET Limited Partnership
Photos by Patrick D. McMakin and Magdalene Bhandngam McMakin
Cover and endpaper design by Magdalene Bhandngam McMakin

ISBN 974-8495-64-7   White Lotus Co., Ltd.; Bangkok
ISBN 1-879155-16-8   White Lotus Co., Ltd.; Cheney

CONTENTS

Acknowledgements vii
Map of Thailand viii
Preface ix
Key to Symbols x
Suggested Places to Visit xi
Garden 1
Mixed Forests 42
Highland Pine Forest 66
Coastal Strand 72
Freshwater Marsh 82
Field 87
Tropical Fruits 104
Index to Species by Family 110
Index of Scientific Names 120
Index of Common Names 128
Index of Thai Plant Names 134
Bibliography 141

Acknowledgements

I would first like to thank my wife Magdalene who helped me with the photography and Thai plant names along with providing constant encouragement. I also received encouragement over the four years in which I gathered data for this guide from Philip H. Moore who originally taught me the pleasure of plant identification. James and Virginia Di Crocco also encouraged me to continue the endeavor and Dr. Tem Smitinand of the Royal Thai Forestry Department's Herbarium assisted me in identifying some of the more obscure species. Note: Any errors are the responsibility of the author. Comments and corrections are invited.

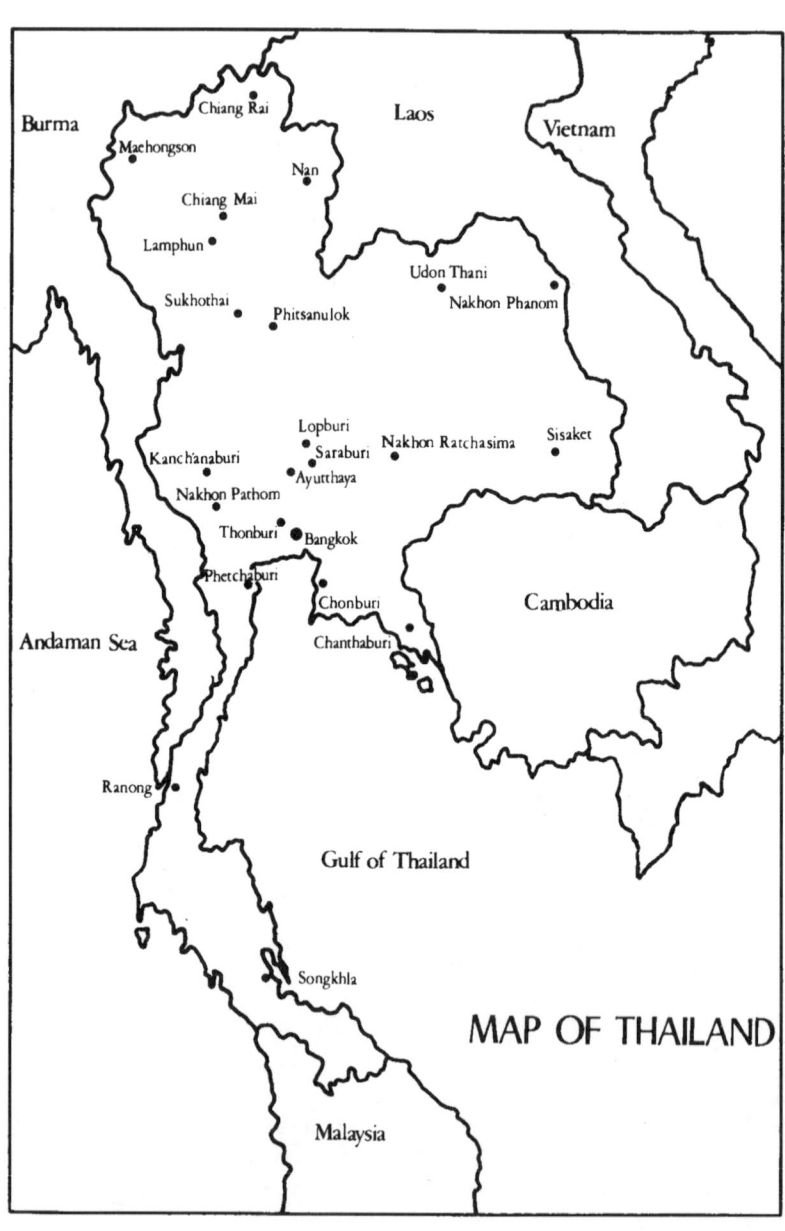

Preface

This field guide has been assembled for the benefit of non-professional botanists or those persons interested in the natural history of Thailand. It is divided into seven separate plant communities, both natural and manmade. We begin with the garden plants and trees which grace parks, residences, temple compounds, embassies and city roadsides. Some of the best places to view garden ornamentals are Nai Lert Park (Hilton Hotel), Lumpini Park, Suan Thonburi, Chulalongkorn University campus, Suan Sampran (Rose Garden) and the nursery cooperatives at Chatuchak and Thevet.

The forests of Thailand include various sub-habitats, some of which are combined for the purposes of this guide since many of the species depicted under Mixed Forests can be found in more than one of the sub-communities which include mixed deciduous forests, tropical ravine forests and dry dipterocarp forests.

Due to its distinct nature, the highland pine forest of high (sub-alpine) elevation has been separated from the above mentioned group. In this community, we find a very thin canopy of tall trees with grassy meadows. The pine in this habitat is *Pinus khasya*. Also oaks within the genera *Lithocarpus*, *Quercus* and *Castanopsis* and some dipterocarps, mainly *Dipterocarpus obtusifolius*, are dominant. The meadows of grass contain few large shrubs, but rather an assortment of herbs which are not found elsewhere in Thailand. Those readers familiar with temperate zone species in Europe and America will recognize many genera in this plant community, including gentians and impatiens. There are also many legumes within the genera *Crotalaria, Desmodium* and *Uraria*.

The coastal strand is the seashore habitat including sandy beaches, rocky shorelines, mangroves and tidal flats. In all of these sub-communities, the flora is salt-tolerant or salt-dependent. The coastal strand habitat extends from the ocean's edge inland to a point where forest or freshwater marsh begins or more often to where man has cleared the land for farms, dwellings and roadways.

Freshwater marshes are those inland wetland areas where aquatic vegetation thrives on land constantly inundated with freshwater. These include floodplains, reed marshes, swamp forests, ditches and canals. Brackish and salt marsh species are included within the coastal strand.

When people think of wildflowers, they often envision fields, roadsides and previously cleared or scrubby areas. Here, relatively dry land herbs dominate and prohibit the return of forest vegetation. Many of these field and waste place flowers are naturalized invader species and common throughout the tropics.

In Thailand's many marketplaces, renowned tropical fruits from different regions in Thailand, rare and curious to visitors, are displayed in colorful profusion. Some of the more intriguing and exotic of these fruits are identified in the final chapter.

Key to Symbols

A. Family
B. Scientific Name
C. Common Name
D. Thai Name
E. Origin or Range (If one country is listed, it is the origin; if more than one country, a region or area is listed, this is the range of the species' natural, uncultivated occurrence.)
F. Location where photographed
G. Description of Species

Suggested Places to Visit

Most of the common garden plants depicted in this guide can be easily found in Bangkok's Lumpini Park, located on Rama IV Road between Wireless and Rajadamri Roads. You may also wish to visit the nursery cooperatives of Thevet on Samsen Road and Chatuchak on Phahonyothin Road near the location of the Weekend Market. For less commonly cultivated species, one should visit Suan Thonburi which is located at the terminus of Air Bus Route 4 near King Mongkut Institute of Technology, Thonburi campus. At Suan Thonburi, most of the plants are labeled with scientific names. A magnificent variety of garden species in a small area can be viewed at Nai Lert Park which comprises the grounds of the Hilton Hotel, located near the intersection of Wireless and New Phetchaburi Roads. Even more obscure species with medicinal uses are cultivated on the Chulalongkorn University campus in the area behind the Community Pharmacy Laboratory, directly opposite Mahboonkrong Center on Phayathai Road.

One of Thailand's remaining natural ravine forests, closest to Bangkok, is Namtok Priew National Park in Chanthaburi where one can see wild gingers, begonias and orchids which require a constantly damp habitat along a forest stream. If one wishes to make a more extended study of a tropical ravine forest in a location with comfortable accommodations, the Author recommends the streamside forest which extends from the Thara Hotel to the mining village of Hot Sompan in the provincial town of Ranong. The ravine forest here is lush and is bordered by mixed deciduous forest, both habitats which contain species unique to South Thailand and Peninsular Malaysia.

Many other forests exist in Thailand which are excellent places for nature observation, such as Khao Yai National Park in Nakhon Ratchasima. Highland pine forests dominate Nam Nao National Park in Phetchabun and Phu Luang and Phu Kradeung National Parks in Loei province. For specific directions and descriptions of Thailand's national parks, one should refer to *Guide to Thailand* by Achille Clarac which is available in major bookstores in Bangkok. The Tourism Authority of Thailand on Rajadamnern Nok Road can provide up-to-date information concerning sights, tours and accommodations in park areas. Wildlife observation trips are also organized by the Siam Society on Soi Asoke (Soi 21, Sukhumvit Road).

Fields and wetlands are to be found all over Thailand. The Author conducted most of his field study of these habitats around the campus of King Mongkut Institute of Technology, Ladkrabang campus. This attractive campus, surrounded by fields and marshes, is best visited by taking a 45-minute train ride from Makkasan Station in Bangkok.

Along Sukhumvit Highway, approaching the provincial city of Chonburi from Bangkok, one can easily explore salt marsh, nipa palm and mangrove habitats next to the road without having to trek through the thick, grey mud which characterizes these plant communities. Good locations in which to view coastal strand species include the many famous beaches of Thailand ranging from Koh Samet, a popular island off the Eastern province of Rayong, to Cha-am Beach in Phetchaburi province and Phuket Island in the South. A one-day trip from Bangkok to a good stretch of coastal strand habitat can be made to the strip of land between the bay and the road between the towns of Bang Saen and Ang Sila in Chonburi province.

Tropical fruit trees are grown in almost every backyard in Thailand as well as in parks and temple compounds Large groves and orchards are to be found in many provinces. Durian orchards are mostly located in Thonburi and Chanthaburi provinces. The most famous pomelos are grown in citrus groves in Nakhon Chaisri district of Nakhon Pathom province. There are large roadside fruit markets in this district. To view longan orchards, one must journey north to Lamphun and Chiang Mai. Large highwayside fruit markets can also be found in Pak Chong district of Nakhon Ratchasima province and in Nakhon Nayok province along Route 33.

For more in-depth scientific research of plant communities by visitors, there are only a few places to find the botanical references listed in the bibliography of this guide, many of which are out of print. They are scattered in various libraries including, most notably, the Chulalongkorn University Main Library Reference Room on Phayathai Road, the Siam Society Library on Soi Asoke, the British Council Library at Siam Square and the Royal Thai Forestry Department Herbarium Library on Phahonyothin Road near Kasetsart University.

1 Adhatoda vasica
2 Asystasia intrusa
3 Barleria cristata

4 Barleria siamensis
5 Barleria lupulina
6 Crossandra undulaefolia

7 Graptophyllum pictum
8 Justicia betonica
9 Justicia gendarrusa

10 Odontonema stricta
11 Pachystachys lutea
12 Pseuderanthemum andersonii

13 Pseuderanthemum carruthersii
14 Pseuderanthemum setricalyx

15 Sanchezia nobilis
16 Thunbergia erecta
17 Thunbergia grandiflora

18 Dracaena hookeriana
20 Crinum amabile
21 Crinum asiaticum

19 Echinodorus cordifolius
22 Eucharis grandiflora
23 Eurycles amboinensis

24 Hymenocallis littoralis
25 Pancratium zeylanicum
26 Polianthes tuberosa

27.1 Zephyranthes candida
27.2 Zephyranthes citrina
27.3 Zephyranthes rosea

28 Artabotrys siamensis
29 Cananga odorata
30 Desmos chinensis

31.1 Allamanda cathartica
31.2 Allamanda cathartica var. williamsii
31.3 Allamanda violacea

32 Cerbera manghas
33 Beaumontia grandiflora
34 Carissa carandas

35 Ervatamia coronaria
36 Holarrhena densiflora
37.1 Nerium indicum

37.2 Nerium oleander
38 Odontadenia speciosa
39.1 Plumeria acutifolia
39.2 Plumeria obtusa

39.3 Plumeria rubra
40 Strophanthus gratus
41 Thevetia peruviana

42 Vallaris glabra
44 Spathiphyllum clevelandii

43 Wrightia religiosa
45 Aristolochia galeata
46 Adenium obesum

47.1 Crescenta alata
47.2 Crescentia alata
48 Crescentia cujete

49 Jacaranda mimosaefolia
50 Millingtonia hortensis
51 Pseudocalymma alliaceum

52 Spathodea campanulata
53.1 Tabebuia chrysantha
53.2 Tabebuia pentaphylla
54 Tecoma stans

55 Bixa orellana
56 Cordia dentata
57 Cordia sebestena

58 Pitcairnea flammea
59 Hydrocleys nymphoides
60.1 Pereskia corrugata

60.2 Pereskia grandiflora
61 Canna generalis
62 Capparis micracantha

63 Cleome speciosa
64.1 Crataeva erythrocarpa
64.2 Crataeva nurvala
65 Lonicera japonica

66 Sambucus canadensis
67 Cochlospermum religiosum

68 Quisqualis indica
69 Terminalia catappa
70 Argyreia nervosa

71 Ipomoea carnea
72 Ipomoea quamoclit
73 Dillenia indica

74 Dillenia suffruticosa
75 Tetracera loureiri
77 Acalypha hispida

76 Rhododendron simsii
78 Acalypha wilkesiana

79.1 Hura crepitans
79.2 Hura crepitans
80 Jatropa curcas
81 Jatropha integerrima

82 Jatropha podagrica
83 Nymphoides aurantiaca
84 Messua ferrea

85 Belacanda chinensis
86 Orthosiphon grandiflorus

87 Couroupita gianensis
88 Acacia auriculaeformis
89.1 Bauhinia acuminata

89.2 Bauhinia purpurea
89.3 Bauhinia tomentosa
89.4 Bauhinia variegata
90 Bauhinia winitii

91 Brownea grandiceps
92 Butea monosperma
93 Caesalpinia coriaria

94.1 Caesalpinia pulcherrima
94.2 Caesalpinia pulcherrima ssp. flava
95.1 Calliandra emarginata

95.2 Calliandra haematocephala
96 Cassia bakeriana
97 Cassia fistula

98 Cassia glauca
99 Cassia spectablis
100 Cassia timoriensis
101 Delonix regia

102 Erythrina crista-galli
103.1 Erythrina indica
103.2 Erythrina indica var. alba
104 Gliricidia sepium

105 Leucaena leucocephala
106 Maniltoa gemmipara
107 Mucuna bennettii
108 Parkinsonia aculeata

109 Peltophorum inerme
110 Phyllocarpus septentrionalis

111 Pterocarpus indicus
112 Samanea saman
113 Saraca bijuga

114 Sesbania grandiflora
115 Tamarindus indica
116 Hemerocallis fulva

117 Buddleia madagascariense
118 Lagerstroemia indica
119 Lagerstroemia speciosa

120 Magnolia coco
121 Michelia longifolia
122 Talauma candollii

123 Galphimia glauca
124 Hiptage benghalensis
125 Malpighia coccigera
126 Stigmaphyllon littorale

127 Tristellateia australasiae
128.1 Abelmoschus moschatus
128.2 Abelmoschus moschatus ssp. tuberosus

129 Althaea rosea
130 Gossypium barbadense
131 Hibiscus moscheutos

132 Hibiscus mutabilis
133.1 Hibiscus rosa-sinensis
133.2 Hibiscus rosa-sinensis

133.3 Hibiscus rosa-sinensis
134.1 Hibiscus schizopetalus
134.2 Hibiscus hybrida

134.3 Hibiscus hybrida
135 Hibiscus syriacus

136.1 Malvaviscus drummondii
136.2 Malvaviscus penduliflorus

137 Thalia geniculata
138.1 Medinilla magnifica

138.2 Medinilla schortechinii
139 Aglaia odorata
140 Melia azedarach

141.1 Ficus altissima
141.2 Ficus benghalensis
141.3 Ficus conglomerata

141.4 Ficus glaberrima
141.5 Ficus religiosa
141.6 Ficus triangularis

142 Musa rosacea
143 Callistemon lanceolatus

144 Eucalyptis sp.
145 Bougainvillea spectabilis
146 Victoria amazonica

147 Ochna integerrima
148 Jasminum sambac

149 Nyctanthes arbor-tristis
150 Fuschsia x hybrida
151 Oxalis rosea

152 Caryota mitis
153 Licuala grandis

154 Papaver sominiferum
155 Passiflora laurifolia
156 Plumbago auriculata

157 Antigonon leptopus
158.1 Clematis sp.
158.2 Clematis sp.

159 Rosa hybrida
160 Anthocephalus cadamba
161 Gardenia jasminoides

162 Hamelia patens
163.1 Ixora chinensis
163.2 Ixora finlaysoniana

163.3 Ixora macrothyrsa
163.4 Ixora stricta
164.1 Mussaenda erythrophylla

164.2 Mussaenda philippica
164.3 Mussaenda philippica var. Queen Sirikit
165 Murraya paniculata

166 Ravenia spectabilis
167 Mimusops elengi

168 Russelia equisetiformis
169 Quassia amara

170 Brunfelsia hopeana
171 Cestrum diurnum
172 Cestrum nocturnum

173 Solandra nitida
174 Pterospermum diversifolium
175 Sterculia foetida

176.1 Heliconia humilis
176.2 Heliconia platystachys
176.3 Heliconia psittacorum

176.4 Heliconia rostrata
177 Schoutenia peregrina
178 Turnera ulmifolia

179 Citharexylum spinosum
180 Clerodendron fragrans
181 Clerodendron petasites

182 Clerodendron quadriloculare
183 Clerodendron spelendens

184 Clerodendron ugandense
185 Duranta repens

186 Gmelina philippensis
187 Petrea volubilis
188 Vitex trifolia

189 Alpinia purpurata
190 Curcuma alismatifolia
191 Curcuma domestica

192 Hedychium coronarium
193 Phaeomeria magnifica

GARDEN

1. A. Acanthaceae B. *Adhatoda vasica*
   D. Sanied E. Native to Thailand
   F. Chulalongkorn University campus, Bangkok
   G. A native forest shrub cultivated for medicinal use and ornamental beauty. It has dark green, entire, glossy, acute, oblanceolate leaves and axillary spikes of white, lobed flowers, tinged with purple in the throat.

2. A. Acanthaceae B. *Asystasia intrusa*
   D. Yah-yah E. S.E. Asia
   F. Ranong
   G. A small, branching herb with ovate-lanceolate leaves. The purple and white flowers are tubular, 5-lobed and occur in racemes at the terminal ends of branches.

3. A. Acanthaceae B. *Barleria cristata*
   C. Philippine Violet D. Angab
   E. India F. Nai Lert Park, Hilton Hotel, Bangkok
   G. A small shrub with opposite, oval leaves pointed at both ends. The flowers emerge from leaf axils and bristly bracts, and are tubular, 5-lobed, solid purple or purple and white.

4. A. Acanthaceae B. *Barleria siamensis*
   C. Siamese Barleria D. Angab
   E. Native to Thailand F. Chulalongkorn University campus, Bangkok
   G. A forest shrub which grows well in gardens, it has opposite, ovate leaves. The flowers are violet and emerge from bristly bracts in leaf axils.

GARDEN

5. A. Acanthaceae   B. *Barleria lupulina*
    C. Hop-headed Barleria D. Salate-pahng-porn
    E. Mauritius   F. Ranong
    G. A small shrub with linear leaves. The cream yellow flowers grow from a spike of green bracts which is tinged with red.

6. A. Acanthaceae   B. *Crossandra undulaefolia*
    C. Dark-leaved Crossandra D. Sahn-goranee
    E. India   F. Bang-pa-in, Ayutthaya
    G. A small, erect shrub with whorled, obovate, acute leaves with acuminate base. The salmon pink or yellow-orange flowers occur in an erect, terminal spike with overlapping green bracts.

7. A. Acanthaceae   B. *Graptophyllum pictum*
    C. Caricature Plant   D. Bai-ngun
    E. Polynesia   F. Chulalongkorn University campus, Bangkok
    G. An erect shrub with oblong-ovate leaves which may be purplish or variegated. The tubular, deep red flowers occur in panicles from leaf axils near the terminal ends of branches.

8. A. Acanthaceae   B. *Justicia betonica*
    C. White Shrimp Plant D. Dtree-chawa
    E. Malaysia   F. Chulalongkorn University campus, Bangkok
    G. A shrub with opposite, ovate-lanceolate leaves, acute at both ends. The small, white flowers occur in long, erect, terminal spikes of conspicuous, green-veined, white bracts.

9. A. Acanthaceae   B. *Justicia gendarrusa*
    D. Kah-gai-daeng   E. India to China and Malaysia
    F. Bangkok
    G. Often used as a hedge shrub, it has erect branches with glabrous, lanceolate leaves and erect spikes of small, white flowers.

10. A. Acanthaceae   B. *Odontonema stricta*
    C. Golden Eranthemum D. Saeng-chan-lek

E. Southern Polynesia F. Bangkok
G. A small shrub with ovate, acute leaves. The flowers occur in axillary spikes. They are pink with five lobes and the throat is spotted with red or purple.

11. A. Acanthaceae B. *Pachystachys lutea*
    C. Lollipop Plant D. Luang-kiriboon
    E. South America F. Dusit Zoo, Bangkok
    G. An erect, herbaceous plant grown in flower beds. The white flowers emerge from the base of an erect spike of golden yellow bracts.

12. A. Acanthaceae B. *Psuederanthemum andersonii*
    D. Khem-muang E. Trinidad
    F. Bangkok
    G. An erect shrub with opposite, elliptic, acute leaves and erect, terminal racemes of lilac-violet blossoms. The flowers are tubular with five lobes.

13. A. Acanthaceae B. *Psuederanthemum carruthersii*
    C. Purple Psuederanthemum D. Bai-nahk
    E. Polynesia F. Bangkok
    G. A shrub with variegated, purplish, opposite leaves. The rosy-purple flowers occur in spike-like racemes.

14. A. Acanthaceae B. *Psuederanthemum setricalyx*
    D. Thong-samrit F. Nai Lert Park, Hilton Hotel, Bangkok
    G. A glabrous shrub with lanceolate leaves and racemes of white, 5-lobed flowers with purplish-red throats.

15. A. Acanthaceae B. *Sanchezia nobilis*
    D. Uang-thong E. S. America
    F. Nai Lert Park, Hilton Hotel, Bangkok
    G. A shrub with broadly-obovate, conspicuously-veined leaves. The tubular, yellow flowers occur in erect, terminal spikes.

16. A. Acanthaceae B. *Thunbergia erecta*
    C. Bush Thunbergia D. Chong-nang

GARDEN

  E. Tropical Africa F. Bang-pa-in Palace, Ayutthaya
  G. Planted as an ornamental shrub for its glossy leaves and dark purple blossoms which are trumpet-shaped with a yellow throat.

17. A. Acanthaceae B. *Thunbergia grandiflora*
  C. Bengal Clockvine D. Sai-inthanin
  E. Native to S.E. Asia F. Bangkok
  G. A climbing vine with opposite, palmately-veined leaves and showy, trumpet-shaped, light violet blossoms which hang in long racemes. It is a native forest plant which is planted in gardens.

18. A. Agavaceae B. *Dracaena hookeriana*
  C. Cape of Good Hope D. Wasana
  E. Tropical Africa F. Bangkok
  G. A shrub with thick, lanceolate leaves and pinkish flowers which are borne in rounded clusters on a long, pendulous panicle.

19. A. Alismataceae B. *Echinodorus cordifolius*
  C. Upright Burhead D. Amae-son
  E. North America F. Suan Thonburi
  G. An erect marsh herb which is cultivated in shallow ponds. The leaves are ovate with an acute apex. The white, 3-petaled flowers occur in clusters at nodes on a long, drooping stalk.

20. A. Amaryllidaceae B. *Crinum amabile*
  C. Red Crinum D. Plub-pleung-daeng
  E. Sumatra F. Bangkok
  G. A large, bulbous plant which grows in damp places, it has large, strap-like leaves and large heads of spidery, white blossoms streaked with pink or maroon.

21. A. Amaryllidaceae B. *Crinum asiaticum*
  C. Crinum D. Plub-pleung
  E. Southern Asia F. Chiang Mai
  G. A large herb with fleshy, strap-like leaves growing from a large, fleshy stalk. The spidery, white flowers have six linear segments. The fruit is round.

Amaryllidaceae

22. A. Amaryllidaceae
    B. *Eucharis grandiflora*
    C. Amazon Lily
    D. Wan-mahachoke
    E. Columbia
    F. Thonburi
    G. An herb with broad, fleshy leaves and a tall bloom spike with striking, white, 6-lobed blossoms with a bell-like center with green throat.

23. A. Amaryllidaceae
    B. *Eurycles amboinensis*
    D. Bua-ngun
    E. Malay Peninsula to N. Australia
    F. Ranong
    G. A fleshy plant growing from a bulb with large, orbicular-cordate leaves. The white, tubular flowers occur in a terminal umbel on a long, erect stalk.

24. A. Amaryllidaceae
    B. *Hymenocallis littoralis*
    C. Spider Lily
    D. Plub-pleung
    E. Tropical America
    F. Bangkok
    G. An herb with fleshy, strap-like leaves which grows from a bulb. The flowers are white with long stamens.

25. A. Amaryllidaceae
    B. *Pancratium zeylanicum*
    D. Wan-nang-lom
    E. Sri Lanka and India
    F. Chiang Mai
    G. An herb with an underground bulb and thick, linear-lanceolate leaves. The white flowers are solitary on an erect spike. The flower is unique in that it has a membranous part connecting the stamens.

26. A. Amaryllidaceae
    B. *Polianthes tuberosa*
    C. Tuberose
    D. Sawn-klin
    E. Mexico
    F. Bang-pa-in Palace, Ayutthaya
    G. An herb with linear leaves and a tall bloom spike of white, waxy flowers. There is a Thai superstition that this plant should not be planted in a home garden as the flowers are associated with funerals and may bring bad luck.

27. A. Amaryllidaceae
    B. *Zephyranthes spp.*
    C. Rain Lily
    D. Bua-farang
    E. S. America and the Caribbean

5

GARDEN

    F. Suan Thonburi (1). Bangkok (2). Suan Sampran, Rose Garden (3)
    G. Low growing, bulbous plants with linear. flat. thick leaves, usually grown as a border along walkways. *Z. candida* (l) has white flowers and is from Chile and Brazil. *Z. citrina* (2) is yellow-flowered and from Guyana. *Z. rosea* (3) has bright pink flowers and is from Cuba.

28. A. Annonaceae   B. *Artabotrys siamensis*
     D. Kradanga-tao   E. Native to Thailand
     F. Suan Sampran, Rose Garden
     G. A small forest tree which is cultivated for its fragrant, yellow-green flowers.

29. A. Annonaceae   B. *Cananga odorata*
     D. Kradanga   E. S.E. Asia
     F. Chiang Mai
     G. A native forest tree which has been widely planted as an ornamental. Its flowers have a strong, but pleasing fragrance. The seeds are multiple, ovoid and glossy green.

30. A. Annonaceae   B. *Desmos chinensis*
     C. Chinese Desmos   D. Sayoot
     E. S.E. Asia   F. Suan Sampran, Rose Garden
     G. A large shrub with thick foliage and very fragrant flowers. The flowers have six, wavy, hanging petals which emerge green, turning yellow and finally orange before dropping off separately.

31. A. Apocynaceae   B. *Allamanda spp.*
     C. Yellow Allamanda (1), William's Allamanda (2), Violet Allamanda (3)
     D. Ban-buree   E. Brazil
     F. Bangkok (1,3), Suan Sampran, Rose Garden (2)
     G. Sprawling shrubs which will climb if given support. Their sap is poisonous. *A. cathartica* (l) has bright yellow, trumpet-shaped flowers. The *var. williamsii* (2) is a double-blossomed variation. *A. violacea* (3) has violet blossoms and is seldom seen.

32. A. Apocynaceae   B. *Cerbera manghas*
     C. Pink-eyed Cerbera   D. Dteen-ped-nam

Apocynaceae

  E. Asia   F. Chulalongkorn University campus, Bangkok
  G. A tree of medium height with milky sap and oblanceolate leaves. The S-petaled flowers are white with a dark center spot. The fruit is ovoid, pendulous and contains poisonous seeds.

33. A. Apocynaceae   B. *Beaumontia grandiflora*
  C. Beaumontia   D. Hi-ran-yiga
  E. N.E. India   F. Nai Lert Park, Hilton Hotel, Bangkok
  G. A climbing shrub with woody stems, milky sap and oblong-ovate leaves. The flowers are bell-shaped and white with a green-tinted center. A similar native species *B. brevituba* can be found in Thai forests.

34. A. Apocynaceae   B. *Carissa carandas*
  C. Natal Plum   D. Nam-daeng
  E. S.E. Asia   F. National Museum, Bangkok
  G. A large, thorny shrub with white sap and star-shaped, white flowers; and an attractive fruit, turning from pink to red when ripe.

35. A. Apocynaceae   B. *Ervatamia coronaria*
  C. Crepe Jasmine   D. Phut-jeep
  E. East Indies   F. Chiang Mai
  G. A medium-sized shrub with pure white flowers. The flowers have five petals which are turned at the edges. One variety has double blossoms.

36. A. Apocynaceae   B. *Holarrhena densiflora*
  C. Jasmine Tree   D. Phut-toong
  F. Suan Thonburi
  G. A shrub with glossy, simple leaves and snow-white, jasmine-like flowers, but it is not a true jasmine species.

37. A. Apocynaceae   B. *Nerium spp.*
  C. Oleander   D. Yee-toh
  E. Asia   F. Bang-pa-in Palace, Ayutthaya (1), Bangkok (2)

G. Large shrubs with linear leaves and poisonous, milky sap. *N. indicum* (1) has red, pink or white flowers which cluster at the ends of branches. *N. oleander* (2) has showier flowers.

38. A. Apocynaceae B. *Odontadenia speciosa*
    D. Ban-buree-saet E. Malaysia
    F. Suan Sampran, Rose Garden
    G. A shrub with glossy, opposite leaves and a fleshy, cream yellow or light orange flower with four petallike lobes.

39. A. Apocynaceae B. *Plumeria spp.*
    C. Plumeria, Frangipani D. Lan-tom
    E. Tropical America F. Chiang Mai (1), Dusit Zoo (2), Suan Thonburi (3)
    G. Small trees with soft wood and no foliage on the lower branches. The leaves of *P. acutifolia* (1) are pointed and the flowers white with yellow centers. The leaves of *P. obtusa* (2) are rounded and waxy and the flowers white. *P. rubra* (3) has red flowers and the same structure as *P. acutifolia*. Some botanists consider them to be the same species. Many Thais consider the plumerias to be sacred and they are commonly seen in temple yards.

40. A. Apocynaceae B. *Strophanthus gratus*
    C. Climbing Oleander D. Ban-toen
    E. West Tropical Africa F. Bangkok
    G. A stout, climbing shrub with thick, opposite, obovate leaves. The trumpet-shaped flowers have five, thick, wavy lobes which are light pink.

41. A. Apocynaceae B. *Thevetia peruviana*
    C. Yellow Oleander D. Rumpoey
    E. Peru F. Nakhon Phanom
    G. A tall shrub or small tree with white, yellow or peach-colored flowers with spiraled petals. It contains a white sap.

42. A. Apocynaceae B. *Vallaris glabra*
    C. Bread Flower D. Chomanat

Bignoniaceae

  E. Java   F. Bangkok
  G. A climbing shrub with simple leaves and fragrant, white, bell-shaped flowers in cymes.

43. A. Apocynaceae   B. *Wrightia religiosa*
  D. Mok   E. Native to Thailand
  F. Suan Sampran, Rose Garden
  G. A medium-sized, sprawling. woody shrub with pendulous, usually double-blossomed flowers. This is a native forest species which has been adapted to gardens.

44. A. Araceae   B. *Spathiphyllum clevelandii*
  C. White Flag   D. Gwuk-mae-chan
  E. Columbia and Venezuela F. Nai Lert Park, Hilton Hotel, Bangkok
  G. An aroid with broad, shiny, dark green leaves. The flower is comprised of a white spadix and spathe on a long, erect stem.

45. A. Aristolochiaceae   B. *Aristolochia galeata*
  C. Rooster-flower   D. Gai-fah
  E. Brazil   F. Chiang Mai
  G. An uncommonly cultivated climbing vine with deeply cordate, ovate leaves. The dark purple-veined blossoms are strangely-shaped being inflated and winged.

46. A. Bignoniaceae   B. *Adenium obesum*
  C. Pink Bignonia   D. Chuan-chom
  E. East Africa   F. Lumpini Park, Bangkok
  G. A popular garden plant in Thailand, it is a succulent shrub with milky sap. The flowers are trumpet-shaped and bright pink.

47. A. Bignoniaceae   B. *Crescentia alata*
  C. Calabash   D. Dteen-ped-farang
  E. Mexico   F. Chulalongkorn University campus (1), Bang-pa-in Palace (2)
  G. This odd tree is comprised of long, drooping branches with no branchlets. Rather, the cross-shaped leaves line the main branch. The

GARDEN

flowers (l) emerge directly from the main trunk and produce a round, green, gourd-like fruit (2).

48. A. Bignoniacea B. *Crescentia cujete*
    C. Calabash D. Nam-dtao-ton
    E. Mexico F. King Mongkut Institute of Technology campus, Lat Krabang
    G. Not as commonly seen as *C. alata*, this species is a fuller shrub with spirally arranged, linear leaves. The flowers are round and bell-shaped and grow from the long, drooping branches. The fruit is a large green gourd which can be dried and used as a container after the pulp and seeds are removed.

49. A. Bignoniaceae B. *Jacaranda mimosaefolia*
    C. Jacaranda D. Sri-trang
    E. Tropical America F. Ranong
    G. A small-medium height tree with opposite, bipinnate leaves with numerous small leaflets. The pale lavender-blue flowers are tubular with five lobes and occur in terminal cymes.

50. A. Bignoniaceae B. *Millingtonia hortensis*
    C. Indian Cork Tree D. Beep
    E. Burma F. Bang-pa-in Palace, Ayutthaya
    G. A favorite tree in Thailand because of its fragrant, white, tubular flowers which grow in profusion on panicles. It is a fairly tall forest tree which is planted in parks.

51. A. Bignoniaceae B. *Pseudocalymma alliaceum*
    C. Garlic Vine D. Gatiem-tao
    E. Brazil and Guyana F. Ranong
    G. A climbing shrub grown on trellises and fences. It has glossy leaves and clusters of purple trumpet-shaped blossoms.

52. A. Bignoniaceae B. *Spathodea campanulata*
    C. African Tulip, Tulip Tree D. Gai-phra-law
    E. Tropical Africa F. Phupin Palace, Chiang Mai

G. A tall tree which is easily toppled by storms due to its soft wood and shallow root system. The scarlet flowers are bell-shaped and crowded in erect clusters at the ends of branches.

53. A. Bignoniaceae B. *Tabebuia spp.*
    C. Yellow Tabebuia (1), Pink Tecoma (2)
    D. Luang India (1), Chompoo-puntip (2)
    E. Tropical America
    F. Chulalongkorn Hospital, Bangkok (1), Lumpini Park, Bangkok (2)
    G. Small-medium height trees with five-foliate leaves with irregular-sized leaflets. The leaves drop off during the dry season when the tubular, crepy flowers bloom. *T. chrysantha* (l) has yellow flowers and the more common *T. pentaphylla* (2) has violet-pink blossoms.

54. A. Bignoniaceae B. *Tecoma stans*
    C. Yellow Elder D. Thong-urai
    E. Tropical America F. Nakhon Ratchasima
    G. A small tree or shrub with dense foliage of opposite. serrate leaflets. The flowers are yellow and tubular-shaped. The seeds are contained in a long, flat pod.

55. A. Bixaceae B. *Bixa orellana*
    C. Lipstick Plant D. Kum-saet
    E. Tropical America F. Chulalongkorn University campus, Bangkok
    G. A small, thickly-foliated tree with alternate, long-stemmed leaves. The flowers are white. The fruit is a red, hairy-spiny pod which contains many small, red seeds. The seeds are an important source of anatto, a food dye of economic importance in some countries such as the Philippines where it is called *achiote*. In Thailand; however, the dye is seldom used and the tree is planted for ornamental purposes.

56. A. Boraginaceae B. *Cordia dentata*
    D. Suwana-pruek E. Mexico
    F. Cha-am Beach, Phetchaburi
    G. A small tree with slightly fuzzy leaves and clusters of crinkly, papery, yellow flowers.

GARDEN

57. A. Boraginaceae B. *Cordia sebestena*
    D. Rampon E. Tropical America
    F. Bangkok
    G. A small-medium height tree with stiff, ovate leaves. The flowers are bright orange and occur in terminal clusters. A similar native species *C. subcordata* can be found in the coastal strand.

58. A. Bromeliaceae B. *Pitcairnea flammea*
    C. Organ Mountain D. Sawn-klin-daeng
    E. Brazil F. Bangkok
    G. A terrestrial herb with flaming red flowers on a tall spike which rises above the lanceolate leaves.

59. A. Butomaceae B. *Hydrocleys nymphoides*
    C. Water Poppy D. Fin-nam
    E. Tropical America F. Dusit Zoo, Bangkok
    G. A perennial, aquatic herb cultivated in shallow ponds. It has a submerged root system and floating, round leaves. The cup-shaped flowers are comprised of three yellow petals.

60. A. Cactaceae B. *Pereskia spp.*
    D. Gulap-maulum-lueng E. Mexico and Brazil
    F. Nai Lert Park, Hilton Hotel, Bangkok (1), Thonburi (2)
    G. Large, climbing shrubs with thick foliage of fleshy leaves and long, black spines. It is often planted as a barrier hedge. *P. corrugata* (l) has orange flowers and shorter spines than *P. grandiflora* (2) which has bright pink blossoms

61. A. Cannaceae B. *Canna generalis*
    C. Canna Lily D. Phutta-raksa
    E. Tropical America F. Bangkok
    G. An erect, perennial, unbranched herb with large, oblong, acute leaves which sheath the stem. The flowers are irregularly-lobed, bright red, yellow and orange and occur terminally on a long raceme. Grows best in wet places. C. indica with thinner-lobed flowers is also seen in Thailand.

Caprifoliaceae

62. A. Capparidaceae B. *Capparis micracantha*
    C. Thai Caper D. Ching-chee
    E. Native to Thailand F. Mae Hong Son
    G. A native forest shrub which is planted as an ornamental It has prickly branches and fuzzy, simple leaves. The white flowers occur in leaf axils of trailing branches. They have four small petals, up to 20 elongated stamens and a terminal ovary on a long stalk.

63. A. Capparidaceae B. *Cleome speciosa*
    C. Spider Flower, Bee Plant D. Pak-sian-farang
    E. Tropical America F. Ranong
    G. A curious-looking, small herb with white, pink or purple—pink blossoms. The lower leaves are six-pointed and palmate. The flowers crown an erect spike which is covered with small leaflets. Seed pods hang suspended from long stems.

64. A. Capparidaceae B. *Crataeva spp.*
    C. Caper Tree D. Kum-bok (1), Kum-nam (2)
    E. Native to Thailand F. Bangkok (1,2)
    G. Medium-height native forest trees which are cultivated as ornamentals. They have terminal clusters of fragrant, white and light yellow flowers. The flowers have four petals, four sepals and up to 20 long stamens emerging from the base of a long ovary stalk. The leaves are trifoliate. The young flowers and leaves are sometimes pickled and the bark used medicinally. *C. erythrocarpa* (1) blooms after most of the leaves have shed and the base leaflets are asymmetrically-veined. *C. nurvala* (2) retains its leaves while blooming and the leaflets are more acuminate and symmetrical than the former. It also grows in a wetter habitat in the wild.

65. A. Caprifoliaceae B. *Lonicera japonica*
    C. Honeysuckle D. Sai-nam-pueng
    E. Japan F. Chiang Mai
    G. A climbing shrub with fragrant flowers which are creamy white or yellow. The leaves are dark green on the upper surface and light green below.

GARDEN

66. A. Caprifoliaceae B. *Sambucus canadensis*
    C. Elder D. Oon-farang
    E. North America F. Doi Sutep, Chiang Mai
    G. A large shrub with pinnately-compound leaves with opposite, lanceolate leaflets. The small white flowers are bunched in large, umbellate corymbs. It can be seen as a garden shrub or where it has escaped at the edges of forests, particularly in highland areas such as Chiang Mai and inland Ranong.

67. A. Cochlospermaceae B. *Cochlospermum religiosum*
    C. Buttercup Tree D. Supanika, Fai-kum
    E. Native to Thailand F. Bangkok
    G. A medium-sized, native forest tree which is planted for its bright yellow, bell-shaped blossoms which bloom in the dry season after most the leaves have shed. A double-blossomed variety can also be seen.

68. A. Combretaceae B. *Quisqualis indica*
    C. Rangoon Creeper D. Leb-mue-nang
    E. India F. Bangkok
    G. A climbing shrub with huge clusters of red, pink and white flowers. It has been found in the wild in Thailand, so may be native or early naturalized.

69. A. Combretaceae B. *Terminalia catappa*
    C. Tropical Almond D. Hu-kwang
    E. Pan-tropical F. KMIT campus, Thonburi
    G. A handsome, medium-sized tree with large, obovate leaves with light green veins. The branches are whorled and horizontal. The leaves often turn red before dropping. The fruit is a green, slightly-winged nut which is edible, but rarely eaten in Thailand. The flowers are small, white and occur on slender spikes.

70. A. Convolvulaceae B. *Argyreia nervosa*
    C. Silver Morning Glory, Elephant Creeper
    D. Bai-la-baht E. India
    F. Nai Lert Park, Hilton Hotel, Bangkok

Dilleniaceae

G. A twining, perennial vine with broad, heart-shaped leaves, the upper surface glabrous and the underside covered with a silvery pubescence. The purple flowers are trumpet-shaped and extend from large, papery-thin bracts.

71. A. Convolvulaceae B. *Ipomoea carnea*
    C. Morning Glory D. Pak-boong-farang
    E. Tropical America F. Suan Sampran, Rose Garden
    G. A cultivated species of climbing vine with many close relatives in Thailand's fields and forests. This species is grown on trellises and fences for its large, light violet blossoms.

72. A. Convolvulaceae B. *Ipomoea quamoclit*
    C. Star Ipomoea D. Sone-gahng-plah
    E. Tropical America F. Bangkok
    G. A climbing vine with fern-like, deeply-lobed leaves and bright red, tubular flowers which open into five lobes which form a star-shaped corolla.

73. A. Dilleniaceae B. *Dillenia indica*
    C. Elephant Apple D. Ma-tad
    E. South and S.E. Asia F. Suan Thonburi
    G. A small tree with large, oblong-lanceolate leaves with dentate margins and pronounced veins. The large, round flower buds look like green fruits, but open into a large, showy, nodding flower. The petals are creamy white, cupping numerous light yellow stamens which are centered by a radiating flowerlike style.

74. A. Dilleniaceae B. *Dillenia suffruticosa*
    C. Malayan Dillenia D. Sahn
    E. Malaysia F. Suan Thonburi
    G. A woody shrub with thick foliage of dentate-margin leaves. The flowers are yellow with five petals and usually bloom singly.

75. A. Dilleniaceae B. *Tetracera loureiri*
    D. Rot-sukon E. Indochina
    F. National Museum grounds, Bangkok

15

GARDEN

G. A woody, climbing shrub with simple leaves with parallel veins. The white, fragrant flowers are small, round filament puffs which occur in cymes.

76. A. Ericaceae B. *Rhododendron simsii*
    C. Chinese Rhododendron D. Gulap-doi
    E. China F. Phupin Palace, Chiang Mai
    G. A showy shrub which grows best in the cool, mountainous areas of the North. Its flowers are an attractive violet-red. Several species of native rhododendrons can be seen in the highland forests of Phu Luang and Phu Kradung in Loei province.

77. A. Euphorbiaceae B. *Acalypha hispida*
    C. Chenille Plant D. Hahng-kra-rok-daeng
    E. India F. Suan Sampran, Rose Garden
    G. An ornamental shrub with large, long-stemmed, broadly-ovate leaves with serrate margins. The bright red, pendulous flowers are comprised of red stamens and look like furry, red cats' tails.

78. A. Euphorbiaceae B. *Acalypha wilkesiana*
    C. Joseph's Coat, Copperleaf D. Hu-ling
    E. Fiji F. Bangkok
    G. Often used as a hedge plant, this shrub has dentate, ovate leaves and spikes of hairy bracts and inconspicuous flowers. There are several varieties including the green and white-leaved (variegated) "Java White", copper-leaved "Macafeana" and a mixed "Tricolor."

79. A. Euphorbiaceae B. *Hura crepitans*
    C. Sandbox Tree, Monkey's Dinner Bell
    D. Po-farang E. West Indies
    F. Chulalongkorn University campus, Bangkok (1,2)
    G. A medium-height tree with a thorny trunk and copious, poisonous sap which can cause blindness. The leaves are ovate, acuminate and prominently-veined. The sub-terminal flower is dark red and cone-shaped (1) and produces a solitary, round, depressed fruit (2) which explodes when ripe.

Guttiferae

80. A. Euphorbiaceae B. *Jatropha curcas*
    C. Physic Nut D. Saboo-dum
    E. Tropical America F. Bangkok
    G. A high shrub with milky sap, lobed leaves and small yellow flowers in axillary or terminal panicles. The fruit is a round, poisonous capsule.

81. A. Euphorbiaceae B. *Jatropha integerrima*
    C. Rose-flowered Jatropha D. Bahtavia
    E. Cuba F. Nai Lert Park, Hilton Hotel, Bangkok
    G. A small tree or shrub with full foliage. The bright red flowers are borne in corymbs and bloom year-round.

82. A. Euphorbiaceae B. *Jatropha podagrica*
    C. Gout Plant D. Hanuman-nung-taen
    E. Central America F. Chulalongkorn University campus, Bangkok
    G. A plant with milky sap, lobed leaves and erect cymes of small red flowers. The stems are often thick in places, hence the common name. The fruit is an ellipsoid capsule.

83. A. Gentianaceae B. *Nymphoides aurantiaca*
    C. Golden Water Snowflake D. Bua-bah
    E. India F. Suan Sampran, Rose Garden
    G. A floating water plant with leaves which resemble those of water lilies, but smaller. The flowers are star-shaped and covered with a hair-like yellow fringe giving them a fuzzy appearance. N. indica with white flowers and larger leaves is also seen cultivated in shallow ponds and large jars.

84. A. Guttiferae B. *Messua ferrea*
    C. Indian Rose Chestnut D. Boon-nahk
    E. S.E. Asia F. Chiang Mai
    G. A large, native forest tree which is grown for its attractive white flowers which have a center tuft of yellow stamens. The flowers

GARDEN

occur singly and sometimes back-to-back from leaf axils or the ends of twigs.

85. A. Iridaceae
    B. *Belacanda chinensis*
    C. Leopard Flower, Blackberry Lily
    D. Wan-hahng-chang
    E. East Asia
    F. Maesa Waterfall Park, Chiang Mai
    G. A lily-like herb which grows from a rhizome and produces attractive orange-red flowers with six lobes

86. A. Labiatae
    B. *Orthosiphon grandiflorus*
    C. Cat's Whiskers
    D. Payup-maek
    F. Suan Thonburi
    G. A small herb with serrate leaves and a tall, erect spike of pure white or violet flowers with long stamens.

87. A. Lecythidaceae
    B. *Couroupita gianensis*
    C. Cannonball Tree
    D. Sa-lah
    E. Guyana
    F. Suan Thonburi
    G. A tall tree which curiously exhibits it showy pink blossoms on snaking branch-like projections which emerge from the main trunk, far below the upper canopy. The large, round, brown fruits rarely appear in Thailand. The tree is considered sacred and can be seen mostly growing in temple yards such as at Wat Bovorn in Bangkok and at Wat Sutthachinda in the city of Nakhon Ratchasima. There are a few trees on the Chulalongkorn University campus.

88. A. Leguminosae
    B. *Acacia auriculaeformis*
    C. Australian Wattle
    D. Kathin-narong
    E. Australia
    F. KMIT campus, Lat Krabang
    G. A small shade tree with flat, blade-like leaves which are actually flattened stems called phyllodes. The flowers hang in thin, brush-like spikes.

89. A. Leguminosae
    B. *Bauhinia spp.*
    C. Orchid Tree
    D. Yotakah (1,3), Chong-ko (2,4)

Leguminosae

    E. Continental Asia (1,3), Tropical America (2) India-Indochina (4)
    F. Si Sa Ket (1), Bangkok (2), Chiang Mai (3), Chulalongkorn University campus (4)
    G. *B. acuminata* (l) is a shrub with butterfly-wing-like (bilobed) leaves and nodding, white blossoms. Similar, but with smaller leaves and yellow blossoms is the seldom seen *B. tomentosa* (3). *B. purpurea* (2) and *B. variegata* (4) are medium-sized trees. The former has purple blossoms and the latter, a native forest species has light pink or white flowers.

90. A. Leguminosae   B. *Bauhinia winitii*
     D. Orapin   E. Endemic to Thailand
     F. Suan Thonburi
     G. A climbing vine which can grow to the top of tall trees. It has a thick, woody stem and a dense crown of small bilobed leaves. The white flower has five petals and grows from the end of a branch in a cluster. The seed pod is wide, flat and turns reddish.

91. A. Leguminosae   B. *Brownea grandiceps*
     C. Scarlet Flame Bean   D. Asoke-sapun
     E. Venezuela   F. Wat Sutthachinda, city of Nakhon Ratchasima
     G. A medium-height shade tree with a finely-shaped, umbrella-like crown of thick foliage. The leaflets are aristate, having an elongated apex. The cluster of bright red-orange flowers hang in a tight ball under the canopy of leaves.

92. A. Leguminosae   B. *Butea monosperma*
     C. Flame-of-the-Forest   D. Thong-gwao
     E. Indo-Malaysia to Laos F. Chulalongkorn University campus, Bangkok
     G. A large, native forest tree with trifoliate leaves with large' rounded leaflets and bright orange-red claw-shaped blossoms which appear during the dry season.

93. A. Leguminosae   B. *Caesalpinia coriaria*
     C. Divi-divi   D. Dunyoeng

GARDEN

    E. West Indies   F. Lumpini Park, Bangkok
    G. A shade tree with compound leaves and pale yellow flowers which cluster in panicles.

94. A. Leguminosae   B. *Caesalpinia pulcherrima*
     C. Pride of Barbados, Dwarf Poinciana   D. Nok-yoong Thai
     E. Tropical America   F. Chulalongkorn University campus (1), Bangkok (2)
     G. A large shrub with bipinnate leaves and red-orange flowers (l) which grow in a circular crown at the ends of branches. A yellow-flowered variety, *ssp. flava* (2) can also be seen.

95. A. Leguminosae   B. *Calliandra spp.*
     C. Red Powder-puff (1), Redhead Powder-puff (2)   D. Poo-daeng (1), Poo-chompoo (2)
     E. Asia   F. Chiang Mai (1), Suan Sampran, Rose Garden (2)
     G. *C. emarginata* (l) is a small shrub with bipinnate leaves. The flowers grow singly from leaf axils and are comprised of a puff of bright red stamens. *C. haematocephala* (2) is a larger shrub with compound leaves and small, white flowers with long, pink stamens which cluster into puff-like blossoms.

96. A. Leguminosae   B. *Cassia bakeriana*
     C. Pink Shower Tree   D. Galapah-pruek
     E. Native to Thailand   F. Ubon Ratchathani
     G. A large forest tree with magnificent clusters of pink-white flowers which bloom during the dry season, it is widely cultivated.

97. A. Leguminosae   B. *Cassia fistula*
     C. Golden Shower   D. Chaiya-pruek, Koon
     E. Tropical Asia   F. Ranong
     G. A medium-height tree with large, compound leaves and pendulous clusters of bright yellow flowers. The seed pods are long and cylindrical.

Leguminosae

98. A. Leguminosae B. *Cassia glauca*
    C. Scrambled Eggs D. Khee-lek-ban
    E. S.E. Asia F. Bangkok
    G. A small tree or shrub growing to only about 10 feet in height. The yellow flowers occur in terminal clusters. The seed pods are flat. It is commonly planted along roadsides in Bangkok.

99. A. Leguminosae B. *Cassia spectabilis*
    C. American Cassia D. Khee-lek American
    E. Central America F. Bangkok
    G. A large tree with spreading branches. The leaves are long and pinnate. The large clusters of golden yellow flowers occur at the ends of branches.

100. A. Leguminosae B. *Cassia timoriensis*
     C. Limestone Cassia D. Khee-lek-lued
     E. Continental South and East Asia F. Chulalongkorn University campus, Bangkok
     G. A tree of medium height with pinnately-compound leaves with oblong leaflets and bright yellow flowers in terminal racemes. It is similar in appearance to *C. glauca*, but a much larger tree.

101. A. Leguminosae B. *Delonix regia*
     C. Royal Poinciana, Flame Tree D. Hahng-nok-yoong
     E. Madagascar F. Bangkok
     G. A large-canopied shade tree with finely-compound leaves and bright red-orange flowers which are streaked with yellow. The seed pod is large, long and flat.

102. A. Leguminosae B. *Erythrina crista-galli*
     C. Cockscomb D. Thong-lang Hongkong
     E. Brazil F. Chulalongkorn University campus, Bangkok
     G. A shrub with trifoliate leaves and bright red, tubular flowers blooming in an erect spike.

GARDEN

103. A. Leguminosae B. *Erythrina indica*
     C. Coral Tree, Tiger's Claw D. Thong-lang
     E. Tropical Asia, Native to Thailand F. Bangkok (1,2)
     G. A stocky, prickly tree with long-stemmed, trifoliate leaves which begin falling in the cold season. The blooms occur terminally on a near-leafless tree and are bright red (1). The claw-like petals partially sheath long stamens. A white-flowered *var. alba* (2) is less commonly seen. The young leaves are edible.

104. A. Leguminosae B. *Gliricidia sepium*
     C. Madre de Cacao D. Khae-farang
     E. Tropical America F. Bangkok
     G. A medium-height tree with compound leaves and dense clusters of flowers which occur in racemes along branches. The pea-like flowers are pink or white with a light yellow spot on the lip.

105. A. Leguminosae B. *Leucaena leucocephala*
     C. Horse Tamarind D. Krathin
     E. Tropical America F. Bangkok
     G. A small tree with finely-divided, light green leaves. The flowers are round, white puffs. The leaf tips and immature pods are popularly used as a vegetable; thus, the tree is often cultivated as a food source.

106. A. Leguminosae B. *Maniltoa gemmipara*
     D. Asoke-khao E. New Guinea
     F. Bangkok
     G. A medium-height tree with glossy, leathery, bilobed leaves. The inflorescence is comprised of drooping clusters of light pink to white, young leaves which cover the small, white flower which produces a one-seeded pod.

107. A. Leguminosae B. *Mucuna bennettii*
     C. Red Jade Vine D. Puang-gomaen
     E. New Guinea F. Bangkok
     G. A large, climbing vine with a woody stem and trifoliate leaves. It produces spectacular, bright red clusters of beak-shaped flowers.

Leguminosae

108.   A. Leguminosae   B. *Parkinsonia aculeata*
      C. Jerusalem Thorn   D. Ratamah
      E. Tropical America   F. National Museum, Chiang Mai
      G. A large, spiny shrub with drooping branches. The long, thin, compound leaves are very peculiar, the leaflets being smaller and wider-spaced than most legumes. The yellow flowers have one reddish-brown petal.

109.   A. Leguminosae   B. *Peltophorum inerme*
      C. Yellow Poinciana, Copper Pod   D. Non-see
      E. S.E. Asia   F. Bangkok
      G. A large shade tree with compound leaves and conical spikes of bright yellow flowers. The seeds are encased in reddish-brown pods.

110.   A. Leguminosae   B. *Phyllocarpus septentrionalis*
      C. Monkey-flower Tree   D. Pradu-daeng
      E. Guatemala   F. Chulalongkorn University campus, Bangkok
      G. A fairly tall tree with compound leaves that shed completely during the dry season when the tree blooms with long, showy, red inflorescence.

111.   A. Leguminosae   B. *Pterocarpus indicus*
      D. Pradu   E. Tropical Asia and Pacific
      F. Bangkok
      G. A medium-large shade tree with compound leaves comprised of large, ovate, acute leaflets. The flowers appear for only a few days each year during the end of the dry season. They are bright yellow, small, papery blossoms clustered in terminal panicles. A single seed is contained in a flat, orbicular, winged pod.

112.   A. Leguminosae   B. *Samanea saman*
      C. Monkeypod, Raintree   D. Jahmjuree
      G. A large shade tree with thick foliage of compound leaves. The flowers are comprised of numerous pink and white stamens.
      F. Lumpini Park, Bangkok

GARDEN

113. A. Leguminosae
 B. *Saraca bijuga*
 D. Asoke-nam
 E. Native to South Thailand and Malaysia
 F. Bangkok
 G. There are a dozen native forest species of Saraca recorded in Thailand, several of which, like *S. bijuga*, have been adapted to gardens as flowering, shade trees. This species is a medium-height tree with glabrous, pinnate leaves, the leaflets oblong-lanceolate and acute. The orange flowers occur in dense corymbs and produce a flat pod.

114. A. Leguminosae
 B. *Seshania grandiflora*
 D. Khae-khao
 E. Tropical Asia
 F. Lumpini Park, Bangkok
 G. A small tree with long, drooping branches and pinnate leaves with oblong-elliptic leaflets. The flowers are white or pink, claw-shaped and edible as are the long, slender pods; thus, the tree is cultivated as both an ornamental and food source.

115. A. Leguminosae
 B. *Tamarindus indica*
 C. Tamarind
 D. Makam
 E. Tropical Africa and Asia
 F. Bangkok
 G. A medium-sized tree with compound leaves. The flowers are light yellow, tinged with pink, and produce a pod with edible pulp.

116. A. Liliaceae
 B. *Hemerocallis fulva*
 C. Day Lily
 E. Europe and Asia
 F. Wat Umong, Chiang Mai
 G. A large herb with glabrous, linear leaves and tawny orange flowers borne on a long spike. The photograph depicts a hybrid, double-blossomed variety. The simple, six-petaled variety is also cultivated in Thailand.

117. A. Loganiaceae
 B. *Buddleia madagascariense*
 D. Rachavadee
 E. Madagascar
 F. Chiang Mai

Magnoliaceae

G. A coarse shrub with pubescent stems, serrate-margin leaves and graceful spikes of small, pure-white flowers.

118. A. Lythraceae B. *Lagerstroemia indica*
C. Crepe Myrtle D. Tabaek
E. S. China F. Bangkok
G. A small tree with small, shiny leaves and terminal clusters of small, pink flowers which appear to be crinkled.

119. A. Lythraceae B. *Lagerstroemia speciosa*
C. Rose of India, Pride of India D. Intanin-nam
E. Native to Thailand and other parts of Asia
F. Bangkok
G. A deciduous forest tree commonly seen in the wild in Thailand and also popularly planted as an ornamental. It has smooth, grey bark and simple, opposite leaves. The crepy, violet-pink blossoms occur in terminal panicles.

120. A. Magnoliaceae B. *Magnolia coco*
C. Magnolia D. Yee-hoob
E. S. China F. Suan Sampran, Rose Garden
G. An ornamental shrub with glossy, oblanceolate leaves, narrowed at both ends. The white flowers are terminal and solitary. The flower is comprised of three greenish sepals and six, waxy, white petals folded into a ball. Unfortunately, this beautiful flower is rarely seen in Thailand's gardens these days.

121. A. Magnoliaceae B. *Michelia longifolia*
C. Champac D. Jum-pee
E. India F. Bang-pa-in Palace, Ayutthaya
G. A medium-sized tree with light green, obovate leaves and waxy, cream-colored blossoms which are beloved for their fragrance. They are often used as tassels for puang malai flower garlands. The orange-flowered *M. champaca* known as *jum-pah* and the small, cream-colored blossoms of *M. fico* known as *jum-pee-kak* are less-seldom seen.

GARDEN

122. A. Magnoliaceae B. *Talauma candollii*
C. Magnolia D. Montha
E. India F. Chiang Mai
G. A small tree with oblanceolate leaves and fragrant, cream-colored flowers. The sepals open when in bloom, but the petals remain folded.

123. A. Malpighiaceae B. *Galphimia glauca*
D. Puang-thong-ton E. Tropical America
F. Bangkok
G. A small, bushy shrub with small, opposite, elliptic-oblong leaves. It is usually planted in flower beds for its year-round, small. bright yellow flowers in erect racemes.

124. A. Malpighiaceae B. *Hiptage benghalensis*
D. No-rah E. India, Malaysia and S. Thailand
F. Chulalongkorn University campus, Bangkok
G. A native, climbing shrub with glossy, ovate-lanceolate leaves, acute or acuminate at the apex. The flowers are fragrant, white with an odd yellow petal and occur in terminal racemes. This plant has medicinal uses.

125. A. Malpighiaceae B. *Malpighia coccigera*
C. Singapore Holly D. Cha Bahtavia
E. West Indies F. Bang-pa-in Palace, Ayutthaya
G. A low shrub with glossy leaves with spiny margins. The flowers are small, white and produce a red berry.

126. A. Malpighiaceae B. *Stigmaphyllon littorale*
C. Orchid Vine, Brazilian Golden Vine E. Tropical America
F. Suan Sampran, Rose Garden
G. A vine with strong, shiny, leathery leaves and canary-yellow flowers with five, crepy petals. A similar species, *S. ciliata*, has heart-shaped leaves.

127. A. Malpighiaceae B. *Tristellateia australasiae*
C. Golden Rod D. Puang-thong-krua

Malvaceae

      E. Australia, Malaysia,   F. Suan Sampran, Rose Garden
          Philippines
      G. A woody, climbing shrub with star-shaped, yellow flowers with red stamens.

128. A. Malvaceae   B. *Abelmoschus moschatus*
     C. Musk Mallow   D. Som-chaba
     E. India to Malaysia   F. Saraburi (1), Suan Thonburi (2)
     G. This species has several variations. One variety (l) has broad, yellow blossoms with a dark crimson center. The flowers turn reddish and fold as the day passes. The leaves are ovate with angular margins. Another variety, *ssp. tuberosus* (2) has deeply and irregularly-lobed leaves with red flowers with white centers. A more common wild variety is depicted with the field plants in this book. Generally, the species is an erect, hairy herb which grows to a height of about one meter. The seeds are musky.

129. A. Malvaceae   B. *Althaea rosea*
     C. Hollyhock   D. Hollyhock
     E. Asia and South Africa   F. Bangkok
     G. An erect, coarse,.annual herb planted in garden plots for its showy blossoms which vary in color. The leaves are fuzzy and lobed.

130. A. Malvaceae   B. *Gossypium barbadense*
     C. Sea Island Cotton   D. Fai-tet
     E. Brazil   F. Bangkok
     G. A branched shrub, 1-3 meters high with purplish stems and 3-5-lobed leaves. The flowers are terminal, yellow turning reddish purple and are seated in a bract with three, toothed lobes. The small, ovoid seeds are covered with fine, white, cotton-like hairs.

131. A. Malvaceae   B. *Hibiscus moscheutos*
     C. Swamp Rose Mallow, Mallow D. Phuttan-rah
        Rose
     E. North America   F. Nakhon Phanom
     G. An erect, coarse herb which is closely related to okra and similar in appearance. The leaves are broad, angular and the flower has five, cream-colored petals with a yellow center.

GARDEN

132.   A. Malvaceae   B. *Hibiscus mutabilis*
      C. Changeable Rose   D. Phuttan
      E. China   F. Saraburi
      G. A small-medium height shrub with broad leaves with irregular margins. The large flowers bloom white in the morning and turn darker pink as the day passes.

133.   A. Malvaceae   B. *Hibiscus rosa-sinensis*
      C. Hibiscus   D. Chaba, Phurahong
      E. Africa   F. Chiang Mai (1), Suan Thonburi (2), Lumpini Park (3)
      G. Woody shrubs which have been hybridized into a wide range of colors. It has glossy, toothed leaves and the flowers are large and showy. They are commonly seen in red (1), pink, orange, yellow, cream and white (2). The double, yellow blossom (3) is a hybrid called "Mist".

134.   A. Malvaceae   B. *Hibiscus schizopetalus*
      C. Coral Hibiscus   D. Phurahong
      E. East Tropical Africa   F. Chiang Mai (1), Ranong (2), Nai Lert Park, Hilton Hotel, Bangkok (3)
      G. This species is commonly seen as a woody, hedge shrub in Thailand's gardens. It is recognized by the flowers with fringed, upturned petals; hanging suspended on long stems. The common variety is red, streaked with white (1). The flower in photograph (2) appears to be a cross between *H. rosa-sinensis* and *H. schizopetalus* and photograph (3) depicts a tricolor hybrid with pink, white and green leaves.

135.   A. Malvaceae   B. *Hibiscus syriacus*
      C. Blue Hibiscus   D. Chaba
      E. India and China   F. Chiang Mai
      G. A small, erect, woody shrub with ovate leaves with dentate margins. The blossoms are violet or blue-violet with crimson centers.

136.   A. Malvaceae   B. *Malvaviscus spp.*
      C. Sleeping Hibiscus, Turk's Cap   D. Chaba

Meliaceae

E. Central America and West Indies
F. Bangkok (1), Chiang Mai (2)
G. M. drummondii (1) is a small, weak shrub with ovate, serrate leaves and red blossoms that appear closed when in full bloom. *M. penduliflorus* (2) is a larger, woody shrub with glossy leaves.

137. A. Marantaceae
B. *Thalia geniculata*
C. Water Canna
D. Kluay-mai-nam
E. Central and South America
F. Nai Lert Park, Hilton Hotel Bangkok
G. An aquatic plant growing in shallow water and rooted in mud. Its broad leaves look much like canna lilies; however, its small, violet flowers hang in pendulous racemes from the terminal ends of tall spikes.

138. A. Melastomaceae
B. *Medinilla spp.*
C. Medinilla
D. Soi-raya
E. Philippines and Java
F. Bangkok (1), Ranong (2)
G. *M. magnifica* (1) is a very showy plant which is not commonly seen outside houses and greenhouses. It can be recognized by its spectacular, pendulous clusters of waxy, pink flowers with pink, wing-like bracts. The broad leaves are deeply-veined. The more commonly seen *M. schortechinii* (2) has smaller leaves and less showy blossoms.

139. A. Meliaceae
B. *Aglaia odorata*
C. Chinese Rice Flower
D. Prayong
E. S.E. Asia
F. National Museum, Bangkok
G. A native forest tree which is cultivated for its showy sprays of golden yellow flowers which cluster in panicles along branches.

140. A. Meliaceae
B. *Melia azedarach*
C. Chinaberry, Indian Lilac, Persian Lilac
D. Lian
E. Tropical Asia
F. Dusit Zoo, Bangkok
G. A small, shrubby tree with bipinnate leaves with toothed leaflets. The flowers are white or pale violet and have dark purple staminal tubes.

GARDEN

141. A. Moraceae B. *Ficus spp.*
     C. Ficus; Bo Tree, Peepul Tree (5) D. Sai (2). Po (5), Sai-bai-sam-liam (6)
     E. Tropical Asia F. Bangkok (all photographs)
     G. *F. altissima* (1) is a large shade tree with leathery leaves and ornamental, ovoid, orange fruits. *F. benghalensis* (2) is also a shade tree with small, glossy leaves and small, yellow fruits. This species has hanging, aerial roots which can eventually root and become support trunks. *F. conglomerata* (3) has clusters of large, reddish fruit which grow from the lower trunk and branches. *F. glaberrima* (4) has abundant clusters of light green fruits growing from main branches. These fruits are favorites of birds. *F. religiosa* (5) is the well-known, sacred tree found in every temple compound. It has acuminate, heart-shaped leaves. *F. triangularis* (6) is a small shrub or tree with leathery, triangular-shaped leaves. All of the above species have inconspicuous flowers.

142. A. Musaceae B. *Musa rosacea*
     C. Lotus Banana D. Kluay-bua
     E. Native to Thailand F. Bangkok
     G. A native herb of damp ravine forests in the North, it grows well in gardens. The pink and yellow flower resembles a lotus bud.

143. A. Myrtaceae B. *Callistemon lanceolatus*
     C. Showy Bottlebrush D. Liu-dok
     E. Australia F. KMIT campus, Lat Krabang
     G. A tall tree with weeping branches festooned with bottlebrush-like flowers comprised of numerous bright red filaments.

144. A. Myrtaceae B. *Eucalyptis sp.*
     C. Gum Tree E. Australia
     F. Ranong
     G. A narrow tree which can eventually reach a height of 40 meters. It has aromatic, alternate, lanceolate, pale green leaves. The white, filament-puff, axillary flowers occur in 5-12-flowered umbels with conical operculum.

Oleaceae

145. A. Nyctaginaceae B. *Bougainvillea spectabilis*
     C. Bougainvillea D. Fueng-fah
     E. Brazil F. Chiang Mai
     G. A woody, thorny, sprawling shrub which will climb a trellis or fence. What appears to be the flowers are actually colorful bracts which may be scarlet, orange or white. The actual flower is small, white and blooms in groups of three.

146. A. Nymphaeaceae B. *Victoria amazonica*
     C. Royal Water Lily D. Bua-kradong
     E. Brazil F. Nai Lert Park, Hilton Hotel, Bangkok
     G. A large, perennial, aquatic herb. Its round, floating leaves are 1-2 meters in diameter. Upturned edges on the lily pads keep them dry on the surface. The unimposing flower is white or pinkish yellow. It can be seen in the moat surrounding Chitralada Palace and at Nai Lert Park.

147. A. Ochnaceae B. *Ochna integerrima*
     C. Mickey Mouse Tree D. Kamlang-changsan
     E. Tropical Africa F. Suan Thonburi
     G. A very attractive, woody shrub which only attains a height of about two meters. Its bright yellow flowers are complimented by the round fruits which are encased in bright red sepals.

148. A. Oleaceae B. *Jasminum sambac*
     C. Jasmine D. Mali
     E. India F. Bang-pa-in, Ayutthaya
     G. A climbing shrub with opposite leaves and terminal clusters of white, fragrant, star-shaped flowers. There are several species of ornamental jasmine in Thai gardens and many native species in forests and the coastal strand.

149. A. Oleaceae B. *Nyctanthes arbor-tristis*
     C. Tree of Sadness D. Kanika
     E. India F. Suan Sampran, Rose Garden

GARDEN

    G. A climbing shrub with clusters of attractive flowers which are comprised of a bright orange tube and five white, petal-like lobes. The flowers bloom at night and drop off in the daytime.

150. A. Onagraceae B. *Fuchsia x hybrida*
 C. Fuchsia E. Central America
 F. Phupin Palace, Chiang Mai
 G. There are many hybrid varieties which are popularly grown in subtropical climates. It grows well in the cooler heights of North Thailand. It is a shrub with dense foliage and spectacular, pendulous flowers. The variety in the photograph is known as "Strawberry Queen".

151. A. Oxalidaceae B. *Oxalis rosea*
 C. Pink Wood Sorrel E. Chile
 F. Phupin Palace, Chiang Mai
 G. A forest herb which is cultivated, in flower beds in partial shade. It grows from a tuberous rootstock . and has radical leaves (growing directly from the root) which are trifoliolate with obcordate leaflets (round with an indented apex). The flowers are purple-pink.

152. A. Palmae B. *Caryota mitis*
 C. Fishtail Palm D. Dtao-chang-goh
 E. Tropical Asia F. Suan Thonburi
 G. A native forest palm of medium height which is commonly planted in gardens. It has bipinnate, triangular-shaped leaves of which the outer edges are ragged.

153. A. Palmae B. *Licuala grandis*
 C. Fan Palm D. Pahm-jeep
 E. Celebes F. National Museum, Bangkok
 G. An ornamental palm with fan-shaped leaves and hanging clusters of bright red berries. There are numerous species of ornamental palms found in Thailand's gardens and Lumpini Park and the Dusit Zoo have sections which are exclusively planted with palms.

154. A. Papaveraceae B. *Papaver somniferum*
 C. Opium Poppy D. Fin

Rosaceae

    E. Asia Minor   F. Chiang Mai
    G. Sometimes cultivated for its showy flowers, the opium poppy is primarily cultivated in cool, highland areas of North Thailand by hilltribe peoples as an annual economic crop. The cup-shaped flowers are red, white or violet on long, erect pedicels. The seed pod yields a milky sap or raw opium. The leaves are coarsely-dentate and clasp the stem.

155. A. Passifloraceae   B. *Passiflora laurifolia*
      C. Passionflower   D. Kratok-rok-farang
      E. Brazil   F. Bangkok
      G. This climbing vine is planted for its purple and white blossoms. The orange fruits are ovoid and are edible, but do not often bear in Thailand.

156. A. Plumbaginaceae   B. *Plumbago auriculata*
      C. Leadwort   D. Payap-mohk
      E. South Africa   F. Bangkok
      G. A shrub which grows to a height of about one meter. It has oblong-obovate leaves and pale blue flowers.

157. A. Polygonaceae   B. *Antigonon leptopus*
      C. Chain-of-Love   D. Puang-chompoo
      E. Mexico   F. Bang-pa-in, Ayutthaya
      G. A vine with tendrils which grows from a starchy, tuberous root. The flowers are bright pink or white and cluster in racemes. It has naturalized in some previously-cultivated areas.

158. A. Ranunculaceae   B. *Clematis spp.*
      C. Clematis   D. Puang-kaeo-kudan
      F. Bangkok (1), Chiang Mai (2)
      G. Climbing shrubs which grow best in cooler climates, they have opposite leaves and axillary flowers comprised of petal-like sepals and numerous white stamens. Species (1) has dark purple-black sepals and oblanceolate leaves. Species (2) has white sepals and trifoliolate leaves.

159. A. Rosaceae   B. *Rosa hybrida*
      C. Rose   D. Gulap

GARDEN

    E. China   F. Doi Sutep, Chiang Mai
    G. Visitors to Thailand are often surprised to find such a dazzling array of roses in gardens such as Suan Sampran and Phupin Palace. The ornamental shrubs have thorns and ovate leaves with serrate margins. They grow well only when pampered with proper irrigation, fertilizer and frequent spraying of pesticides.

160. A. Rubiaceae   B. *Anthocephalus cadamba*
        D. Katum-nam   E. India-S.E. Asia
        F. Suan Thonburi
        G. A fairly large tree with opposite, obovate leaves and round, yellow flowers.

161. A. Rubiaceae   B. *Gardenia jasminoides*
        C. Gardenia   D. Phut-sawn
        E. India   F. Suan Sampran, Rose Garden
        G. A bushy shrub with glossy leaves and large. multi-petaled, very fragrant, white flowers which most often bloom singly. The photograph depicts a variety with variegated leaf.

162. A. Rubiaceae   B. *Hamelia patens*
        C. Scarlet Bush, Fire Bush   D. Phratat Filipine
        F. Nai Lert Park, Hilton Hotel Bangkok
        G. A small shrub with simple, distinctly-veined leaves and weak panicles of bright, orange-red flowers with petals folded into tubular shape.

163. A. Rubiaceae   B. *Ixora spp.*
        C. Ixora   D. Khem, Khem-sethi
        E. Asia   F. Bangkok (1,2,4), KMIT, Lat Krabang (3)
        G. Shrubs with large clusters of red, orange, yellow, pink, cream or white flowers. *I. chinensis* (1) is from China and has red, yellow or hybrid pink blossoms. *I. finlaysoniana* (2) has small, white flowers and can grow into a small tree. *I. macrothyrsa* (3) is a large shrub and has bright red flowers and larger leaves than the other species. *I. stricta* (4) has small leaves and is compact, making it popular for hedges.

Scrophulariaceae

164. A. Rubiaceae B. *Mussaenda spp.*
C. Ashanti Blood (1), Mussaenda (2), Queen Sirikit Mussaenda (3) D. Don-ya (1,2), Don-ya Sirikit (3)
E. Africa (I), Tropical Africa, Asia and the Pacific (2) Philippines (3) F. Chiang Mai (1), Bangkok (2,3)
G. Shrubs with showy flowers, they attain a height of about 3 meters. *M. erythrophylla* (1) has bright scarlet bracts and small, cream-colored flowers. *M. philippica* (2) has large, white bracts and small, bright yellow flowers. *M. philippica var.* Queen Sirikit (3) is a hybrid with pink bracts and yellow flowers.

165. A. Rutaceae B. *Murraya paniculata*
C. Orange Jasmine D. Gaeo
E. Native to Thailand F. Bangkok
G. This is a native forest shrub with small, simple leaves and clusters of fragrant, white flowers with squarish petals.

166. A. Rutaceae B. *Ravenia spectabilis*
C. Pink Ravenia D. Muchalin
E. West Indies F. Suan Sampran, Rose Garden
G. A small shrub with glossy leaves and bright pink, 5-lobed flowers which have a yellow throat.

167. A. Sapotaceae B. *Mimusops elengi*
D. Pikul E. India
F. Suan Thonburi
G. A medium-height shade tree with dense foliage of elliptic leaves. The small, fragrant flowers are cream-colored and grow in axillary clusters; they usually point downward from the undersides of branches. The ellipsoid. reddish-orange fruit is sometimes eaten.

168. A. Scrophulariaceae B. *Russelia equisetiformis*
C. Fire-cracker Flower D. Phratat-chin
E. Mexico F. Suan Thonburi
G. A small shrub with opposite, whorled, scale-like leaves. The drooping flowers occur at the terminal ends of branches, are bright red and tubular.

GARDEN

169.   A. Simarubaceae   B. *Quassia amara*
      C. Bitterwood   D. Phratat-yai
      E. Tropical America   F. Chulalongkorn University campus, Bangkok
      G. A medium-sized shrub with reddish branches and red-veined leaves. The glossy leaves are winged and have five leaflets. The red flower consists of five petals folded into a tube and occur in terminal racemes.

170.   A. Solanaceae   B. *Brunfelsia hopeana*
      C. Yesterday, Today and Tomorrow   D. Phuttachat-sam-see
      E. Brazil   F. Bangkok
      G. A small shrub with tubular flowers which open into five, flat, petal-like lobes. The flowers emerge purple, turning violet and finally white on successive days.

171.   A. Solanaceae   B. *Cestrum diurnum*
      C. China Inkberry   E. Tropical America
      F. Chulalongkorn University campus, Bangkok
      G. A low-growing, many-branched shrub with pale green, oblong leaves. The small, white, tubular flowers grow terminally from leaf axils in racemes. The fruits are pea-sized and contain a purple juice which can be used for ink.

172.   A. Solanaceae   B. *Cestrum nocturnum*
      C. Lady-of-the-Night   D. Ratree
      E. West Indies   F. Chiang Mai
      G. A shrub with pale green, oblong leaves and large sprays of small, white, tubular flowers which have a very strong fragrance at night.

173.   A. Solanaceae   B. *Solandra nitida*
      C. Cup-of-Gold, Golden Chalice   D. Tuay-thong
      E. Mexico   F. Nai Lert Park, Hilton Hotel, Bangkok
      G. A climbing shrub with leathery, obovate leaves and large, yellow, tubular flowers.

Tiliaceae

174. A. Sterculiaceae
     B. *Pterospermum diversifolium*
     D. Champa-tet
     E. India of Malaya and the Philippines
     F. Lumpini Park
     G. A fairly large tree which has alternate, oblong leaves which are light brown on the undersurface. The flowers are cream-colored, fragrant and surrounded by five, thick brown sepals. The ovoid seed pod opens into four symmetrical sections and the seeds are winged.

175. A. Sterculiaceae
     B. *Sterculia foetida*
     D. Sumrong
     E. E. Africa, N. Australia and Tropical Asia
     F. Mae Hong Son
     G. A large, spreading forest tree cultivated for its foliage; it has digitately-compound leaves with 7-9 elliptic-lanceolate, sharply-acuminate leaflets. The flowers have a rank odor and are dull yellow or purplish. They produce a red, obovoid capsule containing black seeds.

176. A. Strelitziaceae
     B. *Heliconia spp.*
     C. Heliconia (I), Hanging Heliconia (2), Japanese Canna (3), Lobster Claw (4)
     D. Dharma-raksa
     E. Tropical America
     F. Maesa Waterfall, Chiang Mai (1), Suan Sampran, Rose Garden (2) Bangkok (3), Dusit Zoo, Bangkok (4)
     G. There are more than a dozen species of heliconia introduced to Thailand. The four most common are shown. *H. humilis* (1) has erect, red bracts tinged with green and small, light green flowers. *H. platystachys* (2) has pendulous, deep red, claw-like bracts and small, yellow flowers. *H. psittacorum* (3) has erect, flaming orange flowers. *H. rostrata* (4) has striking, pendulous, red and yellow bracts enclosing small flowers. All species have broad, lanceolate leaves growing directly from the rootstock.

177. A. Tiliaceae
     B. *Schoutenia peregrina*
     D. Ruang-pueng
     E. Native to Malaysia and S. Thailand

GARDEN

    F. Suan Thonburi
    G. A medium-height forest tree which is planted as a flowering shade tree. The leaves are simple, alternate and the bright yellow flowers occur in axillary clusters. A similar *S. glomerata* can also be seen at Suan Thonburi with densely-flowered cymes with persistent brown calyx and globose ovary.

178. A. Turneraceae   B. *Turnera ulmifolia*
     C. Sage Rose   D. Ban-chao
     E. Mexico   F. Suan Thonburi
     G. An herb which is planted in garden beds, it has serrate leaves and showy cream-colored flowers with a brown-yellow center or yellow flowers with a brighter yellow center. Two varieties are seen in Thailand.

179. A. Verbenaceae   B. *Citharexylum spinosum*
     D. Buknga Bali   E. Barbados
     F. Lumpini Park, Bangkok
     G. A large shrub or small tree with glossy, obovate leaves and drooping spikes of small, star-shaped flowers.

180. A. Verbenaceae   B. *Clerodendron fragrans*
     C. Fragrant Clerodendron   D. Nang-yaem
     E. China   F. Suan Sampran, Rose Garden
     G. A shrub with large, ovate leaves and dense clusters of very fragrant, white flowers which emerge from reddish-tinged calyx-tubes.

181. A. Verbenaceae   B. *Clerodendron petasites*
     D. Kasalong   E. Native to Thailand
     F. Nai Lert Park, Hilton Hotel, Bangkok
     G. A native shrub cultivated for its showy flowers, it has ovate, toothed leaves and long, white, tubular flowers which quickly drop after blooming, leaving a long pistil attached to the ovary.

182. A. Verbenaceae   B. *Clerodendron quadriloculare*
     C. Quezonla   D. Keh-sonla
     E. Philippines   F. Bangkok

Verbenaceae

G. A shrub or small tree with large, opposite, acute leaves which have crenate (wavy) margins. The leaves are dark green above and purple below. The flowers are long, tubular with a pink tube and white lobes and occur in dense, terminal heads.

183. A. Verbenaceae B. *Clerodendron splendens*
  C. Bleeding Heart D. Puang-gaeo-daeng
  E. West Africa F. Thonburi
  G. A climbing, ornamental shrub with glossy, heart-shaped leaves. The flowers are crimson with long, white stamens and emerge from clusters of heart-shaped, white calyxes.

184. A. Verbenaceae B. *Clerodendron ugandense*
  C. Blue Butterfly D. Phee-sua
  E. Tropical E. Africa F. Nai Lert Park, Hilton Hotel, Bangkok
  G. A garden shrub growing to a height of several meters, the elliptic-oblong leaf has a dentate margin and acute apex. The flowers occur in panicles and are five-petaled with four petals light blue and the fifth forming a long, deep purple lip.

185. A. Verbenaceae B. *Duranta repens*
  C. Golden Dewdrop, Pigeonberry D. Puang-muang, Tien-yod
  E. Tropical America F. Bang-pa-in Palace, Ayutthaya
  G. A small-medium height shrub with small, ovate, light green leaves and pendulous, small violet blossoms which produce golden orange berries

186. A. Verbenaceae B. *Gmelina philippensis*
  D. Ching-chai E. Philippines
  F. KMIT, Lat Krabang
  G. A shrub which will climb very high if given support, its long, pendulous flowers are comprised of yellow blossoms which emerge at the end of a tube-like structure of overlapping bracts.

187. A. Verbenaceae B. *Petrea volubilis*
  C. Sandpaper Vine, Purple Wreath D. Chaw-muang, Puang-kram

39

GARDEN

      E. Central America and West Indies   F. Chiang Mai
      G. A woody, climbing, vine with stiff, rough-surfaced leaves and showy, terminal racemes of violet blossoms.

188.   A. Verbenaceae   B. *Vitex trifolia*
      C. Milla   D. Din-saw
      E. Asia   F. Chulalongkorn University campus, Bangkok
      G. A medium-sized shrub with trifoliate leaves, silvery pubescent on the underside. The leaves are aromatic when crushed. The violet flowers are 5-lobed with the lower lobe larger and deeper violet, forming a lip. They occur in terminal panicles. A similar, but prostrate, creeping, native species, *V. ovata*, is depicted in the coastal strand section. It is sometimes cultivated in hanging baskets.

189.   A. Zingiberaceae   B. *Alpinia purpurata*
      C. Red Ginger   D. Khing-daeng
      E. Malaysia   F. Chiang Mai
      G. A large herb growing from a rhizome with broadly lanceolate, glossy leaves and an erect bloom comprised of waxy, bright red bracts and inconspicuous, white flowers.

190.   A. Zingiberaceae   B. *Curcuma alismatifolia*
      C. Lotus Ginger   D. Kah-min-khok
      E. Native to Thailand   F. Chiang Mai
      G. A short, native forest herb cultivated in N. Thailand, it has broadly ovate leaves which sheathe the stem. The erect blossom is comprised of large, bright pink bracts which resemble a lotus blossom and small, actual, purple flowers with yellow throats.

191.   A. Zingiberaceae   B. *Curcuma domestica*
      C. Turmeric   D. Kra-jio
      F. Ranong
      G. An herb with broadly-ovate, plicated leaves which is cultivated for its aromatic rhizomes and showy flowers comprised of erect spikes of fleshy, pink and white bracts. The flowers are sometimes eaten as a vegetable. This herb sometimes spreads into waste places.

Zingiberaceae

192. A. Zingiberaceae B. *Hedychium coronarium*
    C. White Ginger D. Maha-hongse
    E. India F. Ranong
    G. A tall herb with broadly lanceolate leaves growing from rhizomes, it is favored for its white flowers which have a sweet fragrance.

193. A. Zingiberaceae B. *Phaeomeria magnifica*
    C. Torch Ginger D. Gah-lah
    F. Bangkok
    G. A very tall herb with leaves up to 20 ft. in height. The pink, fleshy flower heads rise on separate stalks from the rootstock. The inconspicuous, actual flowers are within a center cone surrounded by a rosette of waxy, pink bracts.

MIXED FORESTS

194.
- A. Acanthaceae
- B. *Barleria prionitis*
- D. Angab
- E. Native to Thailand
- F. Chon Buri
- G. A stiff, erect, branched herb with sharp, axillary spines, three together; entire, acute leaves and terminal spikes of pale orange, lobed flowers, paired and axillary with young leaves.

195.
- A. Acanthaceae
- B. *Clinacanthus siamensis*
- D. Mangkorn-angab-daeng
- E. S. China, Indochina and Thailand
- F. Ranong
- G. A native herb sometimes seen in rural gardens where it is grown for medicinal purposes. It is a weak branching herb with lanceolate leaves and bright red, tubular flowers blooming terminally at branch ends.

196.
- A. Acanthaceae
- B. *Pseuderanthemum graciflorum*
- C. Lilac Eranthemum
- D. Tao-lang-lai
- E. Native to Thailand and Malaysia
- F. Ranong
- G. An erect, forest shrub with opposite, elliptic-lanceolate leaves. The tubular, blue-violet flowers are lobed and occur in erect, terminal racemes. This shrub is very similar to *P. andersonii*, a cultivated ornamental.

197.
- A. Acanthaceae
- B. *Thunbergia fragrans*
- D. Nam-neh-khao
- E. India, Malaysia, Thailand
- F. Koh Samet, Rayong
- G. A twining climber with opposite, ovate-elliptic, glossy leaves and solitary, white, 4-lobed flowers producing a small, sub-globose capsule with four seeds.

Apocynaceae

198. A. Acanthaceae B. *Thunbergia laurifolia*
     C. Thunbergia D. Rahng-juet
     E. Native to Thailand F. Ranong
     G. A woody climber found at the edges of mixed forests. It has oblong leaves with cordate base and acuminate tip. It is distinguished from *T. grandiflora*, another native species commonly cultivated in gardens, which has ovate, pubescent leaves with angled or lobed margins.

199. A. Anacardiaceae B. *Melanorrhoea usitata*
     C. Black Varnish Tree D. Rak
     E. Native to Thailand, Burma and Laos F. Chanthaburi
     G. A medium-height tree found in mixed deciduous forests; it has obovate leaves with a notched apex. The flowers are small, white and occur in terminal panicles. The tree produces a resin used to make lacquer. Some people are allergic to the sap and develop a severe rash.

200. A. Annonaceae B. *Rauwenhoffia siamensis*
     D. Nome-maeo E. Native to Thailand
     F. Chanthaburi
     G. A climbing forest shrub with thick, ovate leaves. The flowers are round and comprised of six, thick, cream-colored petals. The flowers are fragrant and the plant is occasionally cultivated in gardens.

201. A. Annonaceae B. *Uvaria macrophylla*
     D. Nome-chang E. Thailand to Java
     F. Ranong
     G. A woody, climbing shrub with alternate, elliptic-oblong leaves and axillary, dark red flowers producing orange, fleshy fruit.

202. A. Apocynaceae B. *Aganosma marginata*
     D. Maduea-din E. India, Malaysia, Thailand
     F. Ranong
     G. A woody climber with milky sap; glossy, ovate leaves and axillary or terminal cymes of fragrant, tubular-lobed, white flowers.

MIXED FORESTS

203. A. Apocynaceae B. *Wrightea dubia*
     D. Mok-daeng E. Native to Thailand
     F. Ranong
     G. A small tree with opposite, glabrous, obovate leaves with acuminate apex and short, terminal cymes of salver-shaped pink-red flowers.

204. A. Araceae B. *Amorphophallus sp.*
     C. Elephant's Foot Yam D. Buk
     E. Tropical Asia to Fiji and Madagascar F. Ranong
     G. About a dozen similar species of this genera have been recorded in Thailand. They are large forest herbs with underground tubers. The flower is a white or purplish spathe and spadix, sometimes malodorous, and emerges before the solitary leaf appears. The broad, deeply lobed leaf is on a stout petiole which is distinctly mottled. The leaves can he pale green in sunny areas or dark green with a metallic blue sheen in shady ravine forests during the rainy season. The fruit is a globose-to-ovoid, fleshy berry which is white or bluish.

205. A. Araceae B. *Lasia spinosa*
     D. Pak-nam E. S.E. Asia
     F. Bangkok
     G. A perennial herb found growing in moist banks of streams and wet places in forests. The leaves are deeply-lobed with spiny petioles. The erect flower emerges solely from the underground stem and is comprised of a twisting brown spathe around a small, white spadix.

200. A. Araceae B. *Scindapsus siamense*
     D. Hua-chai-naep E. Native to Thailand
     F. Ranong
     G. A vine of wet, ravine forests found climbing on trees by means of its roots. It has heart-shaped leaves which are dark green, blotched with silver and indented on the surface. It is rarely seen to flower.

207. A. Araliaceae B. *Aralia armata*
     D. Glahm-raed E. India to Thailand

194 Barleria prionitis
195 Clinacanthus siamensis
196 Pseuderanthemum graciflorum

197 Thunbergia fragrans
198 Thunbergia laurifolia
199 Melanorrhoea usitata

200 Rauwenhoffia siamensis
201 Uvaria macrophylla
202 Aganosma marginata

203 Wrightea dubia
204 Amorphophallus sp.
205 Lasia spinosa

206 Scindapsus siamense
207 Aralia armata
208 Apama tomentosa

209 Telosma minor
210 Balanophora latisepala
211.1 Begonia rubrovenia

211.2 Begonia thaianum
212 Chloranthus inconspicuus
213 Calycopteris floribunda

214.1 Forrestia griffithii
214.2 Forrestia griffithii
215　Pollia thyrsifolia

216 Wedelia prostrata
217 Cnestis palala
218 Argyreia mollis

219.1 Momordica cochinchinensis
219.2 Momordica cochinchinensis
220.1 Dillenia parviflora

220.2 Dillenia pentagyna
221 Tetracera indica
222 Dioscorea hispida

223.1 Dioscorea pyrifolia
223.2 Dioscorea pyrifolia
224 Shorea roxburghii
225 Elaeocarpus floribundus

226.1 Breynia cernua
226.2 Breynia rhamnoides
227 Homonoia riparia

228 Phyllanthus pulcher
229 Sauropus compressus

230.1 Flagellaria indica
230.2 Flagellaria indica

231 Aeschynanthus marmoratus
232 Curculigo orchoides
233 Scutellaria discolor

234 Leea rubra
235 Acacia catechu
236.1 Acacia farnesiana

236.2 Acacia pennata
237 Adenanthera pavonina
238 Bauhinia bassacensis
239 Bauhinia involucellata

240.1 Bauhinia pottsii var. dicipiens
240.2 Bauhinia pottsii var. velutina
241 Bauhinia pulla

242 Caesalpinia mimosoides
243 Cassia garrettiana
244 Desmodium pulchellum

245 Entada phaseoloides
246 Milletia atropupurea
247 Moghonia strobilifera

248 Parkia speciosa
249 Asparagus racemosus
250 Smilax sp.

251 Dendropthoe sp.
252 Melastoma normale
253 Sonerila deflexa

254 Cyclea peltata
255.1 Broussonetia papyrifera
255.2 Broussonetia papyrifera

256 Ficus benjamina
257 Ficus hirta
258 Ficus racemosa

259 Ardisia crispa
260 Rhodomyrtus tomentosa
261.1 Dendrobium aggregatum

261.2 Dendrobium parishii
261.3 Dendrobium primulinum

261.4 Dendrobium pulchellum
262.1 Paphiopedilum callosum

262.2 Paphiopedilum concolor
262.3 Paphiopedilum exul
262.4 Paphiopedilum parishii

263 Phaius tankervilliae
264 Spathoglottis plicata
265 Aeginetia indica

266 Calamus sp.
267 Licuala spinosa
268 Peperomia pellucida

269 Polygonum barbatum
270 Colubrina asiatica
271 Argostemma sp.

272.1 Ixora sp.
272.2 Ixora sp.
273 Lasianthus oligoneurus
274 Mitragyna rotundifolia

275 Mussaenda sanderiana
276 Pavetta indica

277 Casearia grewiifolia
278 Torenia fournieri
279 Byttneria aspera

280 Helicteres isora
281 Tacca leontopetaloides

282 Colona auriculata
283 Grewia paniculata
284 Clerodendron colebrookianum

285 Clerodendron infortunatum
286 Clerodendron paniculatum

287 Clerodendron spicatum
288 Clerodendron villosum

289 Clerodendron wallichii
290 Congea tomentosa
291 Sphenodesma pentandra

292 Tectona grandis
293.1 Cissus aristata
293.2 Cissus aristata

294 Cissus quadrangularis
295.1 Costus speciosus
295.2 Costus speciosus var. argyrophyllus

296.1 Curcuma sp.
296.2 Curcuma sp.
297.1 Globba leucantha

297.2 Globba obscura
298 Kaempferia pulchra
299 Zingiber spectabile

Begoniaceae

    F. Nam Nao National Park, Phetchabun
    G. A glabrous, prickly forest shrub with tripinnate leaves with ovate-lanceolate, acuminate, serrate leaflets and umbellate panicles of small greenish-white flowers.

208. A. Aristolochiaceae   B. *Apama tomentosa*
      D. Bu-du Bu-lang   E. India to South Thailand
      F. Ranong
      G. A creeping undershrub with thick, ovate, blunt leaves and panicled racemes of brown-violet flowers producing a long, cylindric capsule.

209. A. Asclepiadaceae   B. *Telosma minor*
      C. Tonkin Jasmine   D. Kajorn
      E. Native to Thailand   F. Chon Buri
      G. A climbing vine with opposite, heart-shaped leaves and pendant, axillary umbels of very fragrant, salver-shaped, yellow blossoms. Sometimes cultivated as an ornamental.

210. A. Balanophoraceae   B. *Balanophora latisepala*
      D. Kahk-mahk   E. Indo-Malaysia to Indochina
      F. Nam Nao National Park, Phetchabun
      G. An herb lacking chlorophyll and parasitic on forest tree roots; it has fleshy, cream-colored spikes of white flowers emerging from rhizomes.

211. A. Begoniaceae   B. *Begonia.spp.*
      C. Red-veined Begonia (1),   D. Begonia
          Thai Begonia (2)
      E. Native to Thailand   F. Ranong (1), Chanthaburi (2)
      G. Succulent herbs found in moist, ravine forests, they have fleshy, asymmetrical. indented leaves, pink or white flowers and a winged, thin-walled seed capsule containing numerous small seeds. *B. rubrovenia* (1) has lanceolate leaves and translucent pink stems with red veins. The leaves are red below and green above and may, or may not, he lined with white spots. *B. thaianum* (2) is commonly seen growing epiphytically on damp, mossy rocks. It has broad, light green leaves and light pink flowers.

MIXED FORESTS

212. A. Chloranthaceae B. *Chloranthus inconspicuus*
     D. Foi-fah E. Native to Thailand
     F. Nam Nao National Park
     G. A shrub with opposite, toothed, elliptic-lanceolate leaves with acuminate apex; and terminal, panicled spikes of small, greenish-white flowers producing white berries with black spots.

213. A. Combretaceae B. *Calycopteris floribunda*
     D. Khao-tok-taek E. India, Burma and Thailand
     F. Ranong
     G. A large, woody forest shrub; the young branches and undersides of elliptic-ovate leaves are pubescent. The flowers occur in axillary racemes. The actual flower is small with no petals and is seated within a persistent, five-lobed, greenish-yellow calyx. Sometimes cultivated as an ornamental.

214. A. Commelinaceae B. *Forrestia griffithii*
     C. Griffith's Forrestia E. Thailand and Malaysia
     F. Ranong (1), Chanthaburi (2)
     G. A stout forest herb, sometimes erect, sometimes prostrate, with unbranched stems. The leaves are purplish, hairy on the undersurface and broadly lanceolate. The flowers are in dense heads which occur along the stalk at the nodes. The small flowers are white and emerge from the colorful cluster of sepals which may be reddish-purple (1) or violet (2). There are five species of this genera recorded in the region.

215. A. Commelinaceae B. *Pollia thyrsifolia*
     C. Many-flowered Pollia D. Ra-ruen
     E. Tropical Asia F. Ranong
     G. A forest herb with lanceolate, acuminate leaves and a stiff, snagging pubescence on the stems. The white flowers are borne in a terminal panicle and produce ovoid-globose, blue seed capsules.

216. A. Compositae B. *Wedelia prostrata*
     E. Southern China. probably naturalized in N. Thailand F. Maesa Waterfall Park, Chiang Mai

Dilleniaceae

G. A spreading herb at the edges of forests, it has oblong leaves, narrowed at both ends and obscurely-toothed. The yellow flower heads are only I cm. in diameter.

217. A. Connaraceae B. *Cnestis palala*
 D. Gnoen-gai E. Thailand, Burma, Sumatra and Philippines
 F. Ranong
 G. A climbing shrub found in open places within mixed deciduous forests; with odd-pinnate leaves and racemes of small, pink and white flowers producing a velvety, scarlet-orange, curved fruit.

217. A. Convolvulaceae B. *Argyreia mollis*
 D. Krua-phu-ngoen E. Malaysia
 F. Chon Buri
 G. A woody, climbing vine with milky sap. found at the edges of forests; with alternate, pubescent. heart-shaped, acuminate leaves and campanulate flowers, white outside with a deep purple lip.

219. A. Cucurbitaceae B. *Momordica cochinchinensis*
 C. Spike-fruited Crow's Cucumber D. Fak-khao
 E. Tropical Asia F. Chanthaburi (1,2)
 G. A high-climbing vine with tendrils and deeply-indented, three-lobed leaves. The flower (I) is yellowish-pink or peach-colored. The fruits are edible and were often eaten in former times; they are seldom eaten today. The large, roundish-ovoid fruit (2) is covered with short spines. It turns from green to yellow to orange and bright red successively as it ripens.

220. A. Dilleniaceae B. *Dillenia spp.*
 D. Sahn-hing (1), Sahn-chang (2) E. Native to Thailand
 F. Ranong (1,2)
 G. These are trees of medium height found in mixed deciduous forests. *D. parviflora* (1) blooms with large, five-petaled flowers with a conspicuous, radiating style, after the simple, alternate leaves have dropped in the dry season. *D. pentagyna* (2) has smaller, yellow blossoms and globular fruits. Aside from these two species and the

MIXED FORESTS

two ornamental species depicted in this book, there are six other native species of *Dillenia* found in Thai forests.

221. A. Dilleniaceae B. *Tetracera indica*
      C. Hedge-row D. Tao-orakhon
      E. Malaysia and S. Thailand F. Ranong
      G. A woody, climbing shrub with stiff, toothed, alternate leaves. The flowers are white with many rose-pink stamens in a tuft. The fruit is three-capsuled and beaked and contains small, glossy black seeds with a bright orange-red, hairy appendage.

222. A. Dioscoreaceae B. *Dioscorea hispida*
      C. Wild Yam D. Gloei
      E. Native to Thailand F. Koh Samet, Rayong
      G. A coarse, thorny-stemmed, climbing vine with stiff trifoliolate leaves up to 30 cm. across, with asymmetrical, prominently-nerved, acuminate leaflets. The small, yellowish flowers are pendulous in racemes and produce grey-pubescent, pale green, 3-keeled pods. The tuberous root is reported to be poisonous unless sliced and soaked for a prolonged period of time.

223. A. Dioscoreaceae B. *Dioscorea pyrifolia*
      C. Wild Yam E. Malaysia, Thailand
      F. Koh Samet, Rayong (1,2)
      G. A climbing vine with ovate, shortly-acuminate leaves, some with rounded base, others cordate. The yellow-green flowers are small in long spikes; male (1) and female (2) and produce 3-winged capsules with thin, round lobe-like wings.

224. A. Dipterocarpaceae B. *Shorea roxburghii*
      D. Payohm E. S.E. Asia
      F. Ranong
      G. A deciduous forest tree of medium height; it has grey bark and simple, oblong-elliptic leaves with a shortly-acuminate tip. The white flowers occur in panicles and have a persistent, pinkish, 5-winged calyx and produce a single, round fruit. There are about 20 species of Shorea in Thai forests.

Euphorbiaceae

225. A. Elaeocarpaceae   B. *Elaeocarpus floribundus*
     D. Muen-doi   E. Native to Thailand
     F. Ranong
     G. A large tree found in mixed deciduous forests; it has simple, alternate leaves and white flowers with fringed petals. The flowers occur in racemes. There is a similar species, *E. grandiflorus*, in Thai forests which has similar flowers; however, they occur only several together in leaf axils.

226. A. Euphorbiaceae   B. *Breynia spp.*
     D. Pid-taw (1), Kahng-plah (2)   E. Tropical Asia
     F. Ranong (1), Chanthaburi (2)
     G. These are large shrubs or small, forest trees with simple, alternate, glabrous, thin, elliptic or ovate leaves. The flowers are very small, greenish and axillary. The fruit of *B. cernua* (1) is a globose, red berry seated in a disc-like or saucer-shaped calyx. The fruit of *B. rhamnoides* (2) is a smaller red or pink berry about 5 cm. in diameter.

227. A. Euphorbiaceae   B. Homonoia riparia
     D. Takrai-nam   E. Indo-Malaysia to Indochina
     F. Ranong
     G. A willow-like shrub with alternate, linear-lanceolate leaves and pendulous spikes of small, brown flowers; it grows along forested stream banks.

228. A. Euphorbiaceae   B. *Phyllanthus pulcher*
     D. Wan-torani-san   E. Indonesia to Thailand
     F. Thonburi
     G. A small, erect, unbranching shrub found near streams and often cultivated for medicinal purposes. It has a woody stem with an umbrella-canopy of oval leaflets and small flowers suspended along the leaf stems. The flowers are on long stems and look like tiny red caps.

229. A. Euphorbiaceae   B. *Sauropus compressus*
     D. Yah-hoon-hai   E. Indo-Malaysia and Thailand

MIXED FORESTS

      F. Koh Samet, Rayong
      G. A shrub or small tree with alternate, ovate leaves and axillary, small, male and female flowers producing a globose fruit.

230. A. Flagellariaceae   B. *Flagellaria indica*
      C. False Rattan   D. Wai-ling
      E. Paleotropical   F. Samut Prakan (1), Chanthaburi (2)
      G. A climbing forest vine with tendril-tipped, narrow. alternate leaves. The flowers (l) are creamy white and are clustered in terminal panicles. The fruits (2) are pinkish, hard berries.

231. A. Gesneriaceae   B. *Aeschynanthus marmoratus*
      D. Uang-gnoen-gai   E. Thailand, Malaysia, Burma
      F. Ranong
      G. An epiphytic, pendulous shrub with opposite, ovate, fleshy leaves marbled with purple. The flowers are small, tubular, green.

232. A. Hypoxidaceae   B. *Curculigo orchoides*
      C. Star Grass, Golden-eyed Grass D. Wan-prao
      E. Native to Thailand   F. Maesa Waterfall Park, Chiang Mai
      G. A wet, ravine forest herb which grows from a tuber or corm and has large, lanceolate, plicated leaves and small, yellow, star-shaped flowers emerging from the base of the plant. It is very similar to another species, *Hypoxis aurea*, in this family which has narrower, linear, pubescent leaves and is found in dry, hilly grasslands. Both species were formerly classified as members of the family Amaryllidaceae before being removed to a separate family.

233. A. Labiatae   B. *Scutellaria discolor*
      C. False Violet   E. India, S.W. China, S.E. Asia
      F. Ranong
      G. A small forest herb found along stream banks; the leaves are ovate with rounded apex and arranged in a basal rosette. The flower is comprised of a violet corolla tube with a lighter lip and occurs on an erect, leafless spike.

Leguminosae

234. A. Leeaceae
     B. *Leea rubra*
     D. Kah-dung-bai
     E. Native to Thailand
     F. Chanthaburi
     G. An erect shrub which is often found in low, wet areas of mixed deciduous forest; it has pinnately-compound leaves with simple, slightly-serrate leaflets. The flowers are blood red and clustered in corymbs. The fruits are clusters of light green berries.

235. A. Leguminosae
     B. *Acacia catechu*
     C. Kathe of India, Catechu Tree
     D. See-siad
     E. Native to Thailand
     F. Chokchai Ranch, Pak Chong, Nakhon Ratchasima
     G. A medium-height forest tree which is sometimes planted as a handsome, roadside tree. The leaves are finely compound and fold quickly when injured. The stems are thorny. The white flowers are comprised of filament-tufts on a long, cylindrical spike from leaf axils. The seeds are contained in brown, flat pods which have wavy margins. The heartwood of the tree yields a gum-resin (catechu) which can be used for dying and tanning.

236. A. Leguminosae
     B. *Acacia spp.*
     D. Krathin-tet (1), Nam-ki-ret (2)
     E. Tropical America (1), Tropical Africa (2) Both species have naturalized in Thailand
     F. Ranong (1), Saraburi (2)
     G. Thorny, forest lianas which will clump into shrubby thickets at the edges of forests and streams. They have finely-compound, bipinnate leaves and round, filament-puff flowers. *A. farnesiana* (l) has yellowish flowers and *A. pennata* (2) has white flowers and the immature flower buds are reddish.

237. A. Leguminosae
     B. *Adenanthera pavonina*
     C. Sandalwood Tree
     D. Maklam-tah-chang
     E. S.E. Asia
     F. Wat Umong, Chiang Mai
     G. A large, forest tree which can be cultivated as a handsome shade tree; it has bipinnate leaves with elliptic leaflets. The flowers are small, yellow and numerous on a spike-like raceme.

MIXED FORESTS

238. A. Leguminosae
    B. *Bauhinia bassacensis*
    D. Chong-ko
    E. Thailand and Indochina
    F. Chon Buri
    G. A woody, climbing shrub with bilobed leaves and greenish-white flowers on corymbose racemes.

239. A. Leguminosae
    B. *Bauhinia involucellata*
    D. Salaeng-pan
    E. S.E. Asia
    F. Phetchabun
    G. A high-climbing, woody shrub with bilobed leaves and terminal racemes of white flowers with five, crepy lobes. The pod is long and flat.

240. A. Leguminosae
    B. *Bauhinia pottsii*
    D. Chong-ko Taksin
    E. Native to Thailand
    F. Trat (1), Ranong (2)
    G. Shrubs with long, drooping branches and bilobed leaves, the flowers occur in short racemes terminally or along branches. The flowers of *var. dicipiens* (1) are crepy and white with a yellow spot. *Var. velutina* (2) has silvery-sheened leaves and flowers with narrow, red-orange and yellow petals.

241. A. Leguminosae
    B. *Bauhinia pulla*
    C. Thai Bauhinia
    D. Salaeng-pan-tao
    E. Native to Thailand
    F. Saraburi
    G. A woody, climbing vine which will clump into a large shrub at the edges of forests; it has bilobed leaves and greenish-white flowers in erect racemes along the stem.

242. A. Leguminosae
    B. *Caesalpinia mimosoides*
    D. Pak-pu-yah
    E. Native to Thailand
    F. Chanthaburi
    G. A large shrub or small tree found in mixed forest and scrubby areas; it has finely-compound leaves and numerous hooked thorns on the stems. The flowers are yellow and cluster terminally at branch ends.

243. A. Leguminosae
    B. *Cassia garrettiana*
    D. Sa-meh-san
    E. Native to Thailand

Leguminosae

F. Chon Buri
G. An evergreen tree with pinnate leaves with acute, ovate leaflets and terminal panicles of bright yellow flowers.

244. A. Leguminosae B. *Desmodium pulchellum*
C. Bracteate Desmodium D. Klet-plachon
E. Malaysia and Thailand F. Nam Nao National Park, Phetchabun
G. A shrub with trifoliate leaves, the basal leaflets reduced, the larger obovate. The small, white flowers are concealed in pendulous racemes of overlapping, leafy, green bracts.

245. A. Leguminosae B. *Entada phaseoloides*
D. Saba E. Native to Thailand
F. Namtok Priew National Park, Chanthaburi
G. A woody forest liana which climbs to great heights; the main stems are twisted or angularly-spiralled. It has bipinnate leaves and the white flowers are crowded on a long, axillary spike from the upper leaves. The seed pods can be over one meter in length, are very woody and jointed between seeds.

246. A. Leguminosae B. *Milletia atropupurea*
D. Kasae E. Indo-Malaysia and S. Thailand
F. Ranong
G. A medium-height forest tree with glossy, compound leaves with narrow, oblong leaflets. During full bloom, the tree is covered with terminal racemes of dark red blossoms with claw-shaped petals. The falling petals carpet the ground beneath the tree.

247. A. Leguminosae B. *Moghonia strobilifera*
D. Khee-daeng E. Southern and S.E. Asia
F. Ranong
G. An erect, branched shrub with velvety branches and oblong to ovate, slightly acuminate leaves. The small flowers occur in terminal racemes of persistent bracts.

248. A. Leguminosae B. *Parkia speciosa*

53

MIXED FORESTS

    D. Sataw
    E. Native to S. Thailand and Malaysia
    F. Ranong
    G. A large, forest tree with a spreading crown of dense, compound leaves; its long, green seed pods yield seeds which look similar to shelled Western lima beans, but are crisp and constitute a favored, cooked Thai vegetable. The trees are most commonly propagated in South Thailand.

249. A. Liliaceae   B. *Asparagus racemosus*
    C. Wild Asparagus Fern   D. Sam-sip
    E. Tropical Asia and Africa   F. Chon Buri
    G. A scandent, thorny shrub with minute, linear leaves and small, white flowers in racemes. The fruit is a small berry.

250. A. Liliaceae   B. *Smilax sp.*
    D. Tao-wun-yung   E. Tropical Asia
    F. Chanthaburi
    G. A woody, climbing vine with tendrils and stiff, oblong, slightly acuminate, prominently 3-nerved leaves. The flowers are very small in peduncled umbels. Six species of *Smilax* have been recorded in Thailand.

251. A. Loranthaceae   B. *Dendrophthoe sp.*
    D. Kah-fak   E. Native to Thailand
    F. Bang-pa-in, Ayutthaya
    G. A glabrous, parasitic shrub with thick, leathery, ovate leaves and small, axillary clusters of yellow flowers with 5 thin lobes and a globose ovary.

252. A. Melastomaceae   B. *Melastoma normale*
    C. Straits Rhododendron   D. Ah-luang
    E. Native to Thailand   F. Ranong
    G. An evergreen shrub which is found in sunny areas of open forest. It reaches a height of several meters. It is recognized by reddish-tinged stems and prominently-3-nerved, oblong, acute leaves and showy, purple, 5-petaled flowers.

Moraceae

253. A. Melastomaceae B. *Sonerila deflexa*
 D. Sao-sanom E. Native to Thailand
 F. Ranong
 G. A small, forest herb with simple, opposite, ovate leaves sometimes lined with white spots. The flowers are violet with three lobes and three yellow stamens. The fruit is a three-sided, elongated capsule.

254. A. Menispermaceae B. *Cyclea peltata*
 E. Thailand, Burma, Cambodia, F. Trat
 Vietnam
 G. A slender, vine-like, climbing shrub with peltate, ovate leaves and cream-colored, small flowers in axillary racemes.

255. A. Moraceae B. *Broussonetia papyrifera*
 C. Paper Mulberry D. Poh-krasah
 E. Native to Thailand F. Saraburi
 G. A small tree, finely pubescent with large, opposite, orbicular-ovate, acute-acuminate apex, cordate, toothed, irregularly-lobed leaves. The male flowers (I) are small, greenish-yellow in pendulous spikes from leaf axils. The fruit (2) is globose, orange and comprised of juicy phalanges.

256. A. Moraceae B. *Ficus benjamina*
 D. Sai-yoi-bai-laem E. South and S.E. Asia
 F. Ranong
 G. This forest species begins as an epiphytic seedling and becomes a large tree with a thick trunk and broad crown. It has simple, latex-bearing, ovate leaves. The fruits grow along branches and are colorful berries that turn from green to yellow and red as they ripen.

257. A. Moraceae B. *Ficus hirta*
 D. Maduea-khon E. Native to Thailand
 F. Trat
 G. A small, forest shrub which only reaches a height of two meters. It has very hairy branches and leaves. The leaves are alternate and trilobed. The small, fuzzy, berry-like fruits are yellow-orange or red when ripe and are borne close to the stem.

MIXED FORESTS

258. A. Moraceae B. *Ficus racemosa*
     D. Maduea-uthumpon E. Native to Thailand
     F. Chanthaburi
     G. A large tree with simple, alternate, ovate leaves. The small flowers occur in short racemes emerging from the trunk and lower branches and produce clusters of globose berries.

259. A. Myrsinaceae B. *Ardisia crispa*
     C. Village Ardisia D. Jum-kong
     E. Thailand and Malaysia F. Ranong
     G. A small, woody, forest shrub with narrow, slightly toothed leaves and waxy, pink flowers with overlapping petals, occurring in drooping umbels from leaf axils. The fruit is a bright red, one-seeded berry.

260. A. Myrtaceae B. *Rhodomyrtus tomentosa*
     C. Rose Myrtle D. Tu-pruet
     E. China to India and S.E. Asia F. Songkhla
     G. A low-growing shrub with glabrous, elliptic-obovate, blunt, three-nerved leaves. The showy flowers are axillary, pink and 5-petaled.

261. A. Orchidaceae B. *Dendrobium spp.*
     D. Uang-pueng (1), Uang-atakrit (2), Uang-sai-prasat (3) Uang-chang-nao (4) E. Native to Thailand
     F. Chiang Mai (1,4), Bangkok (2), Ranong (3)
     G. There are many species of *Dendrobium* in Thai forests, only a few of the most common which are listed here. They are epiphytic plants, growing on trees, with bulbous, usually elongated stems and racemose flowers. *D. aggregatum* (1) has yellow blossoms with a broad, darker yellow lip. *D. parishii* (2) has deep purple blossoms with a fuzzy lip. *D. primulinum* (3) is light violet-pink with a large, light yellow lip. *D. pulchellum* (4) has pink blossoms with two maroon spots within a very shaggy lip.

262. A. Orchidaceae B. *Paphiopedilum spp.*
     C. Lady's Slipper D. Uang-kankoke (1), Rongtao-naree-rueng-prachin (2), Rongtao-naree (3), Rongtao-naree-muang-kan (4)

Orobanchaceae

E. Native to Thailand
F. All photos were taken of cultivated specimens.
G. The lady slippers are terrestrial orchids, usually found growing in damp leaf mold. They generally have lanceolate or oblanceolate leaves and 1-3 flowers on an erect stalk. The genera is easily recognized by the flower's sac-like lip. Eight species have been recorded in Thailand, the most common of which are depicted here. *P. callosum* (1) has a dark purple sac. *P. concolor* (2) has a velvety, creamy yellow blossom. *P. exul* (3) has lanceolate leaves and a glossy, yellow, purple-specked blossom. *P. parishii* (4) has a green blossom, tinged with violet and characterized by long, hanging, sepals.

263. A. Orchidaceae
B. *Phaius tankervilliae*
D. Uang-prao
E. Indo-Pacific and Thailand
F. Chiang Mai
G. A tall, terrestrial orchid with psuedobulbs and lanceolate leaves. The erect raceme has reddish-brown flowers with a purple-tinged lip.

264. A. Orchidaceae
B. Spathoglottis plicata
C. Philippine Ground Orchid
D. Uang-din
E. Indo-Pacific, Native to Thailand
F. Ranong
G. A tall, terrestrial orchid with plicated leaves and round psuedobulbs. The small flowers are rose lavender on an erect stalk and produce cylindrical pods. Found at the edges of forests in sunlight. A related species, *S. eburnea*, has white flowers with a yellow lip and is found in highland pine forests.

265. A. Orobanchaceae
B. Aeginetia indica
C. Broomrape
D. Yah-khao-gum
E. India to S.E. Asia
F. Nam Nao National Park, Phetchabun
G. A fleshy, unbranched, erect parasitic herb without chlorophyll. The tubular, purple flower emerges from a large, nodding, sleeve-like calyx at the terminal end of the stem which contains minute leaf scales.

MIXED FORESTS

266. A. Palmae B. *Calamus sp.*
      C. Rataan Palm D. Wai
      E. Tropical Asia F. Chanthaburi
      G. Of 175 species of rataan in Tropical Asia, a dozen have been recorded in Thailand. They are slender, very spiny, often climbing, less often erect palms with alternate, pinnate leaves. The core of the main stem is edible in most species.

267. A. Palmae B. *Licuala spinosa*
      C. Fan Palm D. Kraproh
      E. Native to Thailand F. Trat
      G. A small palm, it is recognized by its round leaves comprised of radiating, plicated leaflets with toothed margins. The flowers occur on long spike-like racemes from the center of the palm.

268. A. Piperaceae B. *Peperomia pellucida*
      C. Silverbush D. Pak-kra-sank
      E. Native to Thailand F. Ranong
      G. A succulent herb which is found in damp areas of shady forest and appears as a weed in urban gardens. It has translucent stems and heart-shaped leaves. The flowers are tiny and borne on a light green spike.

269. A. Polygonaceae B. *Polygonum barbatum*
      C. Knotweed D. Pak-pai-nam
      E. Tropical Asia F. Chiang Mai
      G. An herb found at the edges of forests in wet places. It has slightly pubescent, lanceolate leaves. The flowers are pink, small and clustered in slender, spike-like racemes which are usually nodding.

270. A. Rhamnaceae B. *Colubrina asiatica*
      D. Kan-soeng E. Paleotropical
      F. Rayong
      G. A climbing shrub with trailing branches of alternate, shiny, dentate, acuminate, slightly cordate, ovate leaves. The flowers are small, yellowish green, axillary and produce a green-to-brown, pea-sized fruit. Can also be found along the seashore

Rubiaceae

271. A. Rubiaceae B. *Argrostemma sp.*
 D. Lin-kuram E. Native to Thailand
 F. Chanthaburi
 G. A small herb with fleshy, basal, ovate-lanceolate leaves and erect, peduncled cymes of white, star-shaped flowers. Found growing on mossy rocks in ravine forests. There are many species of this genus in Thailand and Malaysia.

272. A. Rubiaceae B. *Ixora spp.*
 C. Ixora D. Khem
 E. Native to Thailand F. Ranong (1,2)
 G. Over two dozen species of Ixora have been recorded in Thailand, about half a dozen of which are introduced ornamentals. The native species are found in mixed deciduous and ravine forests. Species (l) has large, obovate, dark green leaves and orange-red flowers. Species (2) has red flowers and light green, lanceolate leaves. All are shrubs with woody branches or very small trees with flowers in terminal umbels producing red or green berries turning black when ripe.

273. A. Rubiaceae B. *Lasianthus oligoneurus*
 E. Native to Thailand F. Nam Nao National Park, Phetchabun
 G. A glabrous shrub with dark green, oblong leaves and small, axillary flowers. There are 15 species of this genera recorded in Thai forests. This species is conspicuous because of its bright blue, berry-like fruits.

274. A. Rubiaceae B. *Mitragyna rotundifolia*
 D. Kratoom-nern E. India to Thailand and the Philippines
 F. Bangkok
 G. A sprawling, woody, shrub or small tree with elliptic-ovate leaves and cream-colored, fragrant flower heads which consist of a puff-ball of small petals and stamens.

275. A. Rubiaceae B. *Mussaenda sanderiana*
 D. Khem-khao E. Native to Thailand

MIXED FORESTS

    F. Ranong
    G. A shrub with drooping branches and simple, opposite, fuzzy leaves. The flowers are small, orange and tubular with five lobes. They occur several together at branch ends and are surrounded by large, white. leafy bracts.

276. A. Rubiaceae   B. *Pavetta indica*
       D. Khem-khao   F. Ranong
       G. An erect, glabrous shrub with elliptic-oblong, acuminate-cuneate leaves with about 8 prominent nerves on each side of the midrib. The flowers are slender, white, tubular with four lobes and occur in many-flowered, terminal clusters.

277. A. Samydaceae   B. *Casearia grewiifolia*
       D. Kruay   E. Malaysia, Thailand and Cambodia
    F. Wat Umong, Chiang Mai
    G. A large forest tree with a spreading crown and alternate, oblong leaves with acute apex and small flowers-in heads. It is recognized by its large, fleshy, bright orange fruit.

278. A. Scrophulariaceae   B. *Torenia fournieri*
       C. Wishbone Flower   D. Waeo-mayurah
       E. Thailand and Indochina   F. Ranong
       G. An erect, annual herb with opposite, ovate leaves and quadrangular stems. The small flowers occur, often singularly, terminally and are tubular with bluish-purple corolla lobes and a yellow spot in the throat. There are three lower lobes and an upper concave lip. This herb can also be cultivated as an ornamental in flower beds.

279. A. Sterculiaceae   B. *Byttneria aspera*
       D. Tao-thong-thuean   E. Malaysia and S. Thailand
    F. Ranong
    G. A climbing shrub with lobed, broadly-ovate leaves and small flowers in axillary cymes; it is recognized by its reddish, spiny, round, pendulous fruits.

Verbenaceae

280. A. Sterculiaceae B. *Helicteres isora*
     D. Paw-bit E. Native to Thailand
     F. Chon Buri
     G. A coarse, erect, slightly-branched shrub with pubescent, alternate, ovate leaves, deeply-toothed at the apex. The orange-red flowers have two large and two small lobes and an elongated pistil. They occur in axillary clusters, short panicles along the length of branches.

281. A. Taccaceae B. *Tacca leontopetaloides*
     C. Polynesian Arrowroot D. Singto-dum, Mai-thao-rusee
     E. Tropical Asia and the Pacific F. Chon Buri
     G. A large herb most often found in open forest and the coastal strand; it has a tuberous root which is poisonous unless processed to render its edible starch. The leaves are tripartite and deeply, irregularly-lobed. They are spreading, up to 1.5 meters across. The green flowers are crowded terminally on a long, erect spike and have filiform bracts.

282. A. Tiliaceae B. *Colona auriculata*
     D. Po-pahn E. Native to Thailand
     F. Chanthaburi
     G. A large shrub or small tree with alternate, toothed, oblanceolate leaves and axillary, yellow flowers.

283. A. Tiliaceae B. *Grewia paniculata*
     D. Mah-kome E. Tropical Asia
     F. Chiang Mai
     G. A shrub or mixed deciduous forests, it has coarse, pubescent, oblong-ovate leaves which are prominently 3-nerved at the base. The young leaves are glossy and reddish. The flowers are yellow, 5-lobed and occur in terminal panicles. The fruit is a black berry.

284. A. Verbenaceae B. *Clerodendron colebrookianum*
     D. Ping-khao E. India to Thailand
     F. Nam Nao National Park
     G. A shrub with broadly-ovate, cordate leaves and dense heads of star-shaped, white blossoms with long stamens. Sometimes cultivated as an ornamental.

MIXED FORESTS

285.   A. Verbenaceae   B. *Clerodendron infortunatum*
       C. Indonesian Clerodendron   D. Pin-daeng
       E. Indonesia, probably naturalized   F. Ranong
           in Thailand
       G. An erect shrub found at the edges of mixed forests, particularly in South Thailand; it has ovate leaves and terminal panicles of white flowers with a red center and long, white stamens; and a red calyx.

286.   A. Verbenaceae   B. *Clerodendron paniculatum*
       C. Pagoda Flower   D. Chat-fah
       E. Native to Thailand   F. Chanthaburi
       G. A leafy shrub in areas of open forest with broadly-ovate, shiny leaves with cordate base and 3-7 angularly-lobed margins. The bright red flowers are borne in terminal, erect-pyramidal panicles.

287.   A. Verbenaceae   B. *Clerodendron spicatum*
       D. Yah-nuat-suea   E. Native to Thailand
       F. Chiang Rai
       G. A ravine forest shrub which reaches a height of about two meters; it has opposite, oblong leaves, acute at both ends. The pink flowers are borne on an erect, terminal, tapering panicle comprised of pinkish branchlets or stems.

288.   A. Verbenaceae   B. *Clerodendron villosum*
       D. Nang-yaem-pah   E. Native to Thailand
       F. Ranong
       G. A shrub found at the edges of evergreen forests (ravines); it has quadrangular branches and stems; and ovate, opposite, hairy leaves. The white flowers are borne in an erect, terminal panicle.

289.   A. Verbenaceae   B. *Clerodendron wallichii*
       D. Raya-khao   E. Native to Thailand
       F. Ranong
       G. A shrub found in hilly forest areas; it has glabrous, linear-oblong leaves. The white flowers have a conspicuous reddish calyx and occur in terminal panicles.

Vitaceae

290. A. Verbenaceae B. *Congea tomentosa*
     C. Shower-of-Orchids D. Puang-pradit
     E. Malaysia, Burma and South F. Ranong
        Thailand
     G. A climbing shrub with spectacular, long, trailing clusters of purple-pink blossoms which are actually comprised of velvety bracts at the base of tiny, white, tubular flowers. It is often cultivated as an ornamental.

291. A. Verbenaceae B. *Sphenodesma pentandra*
     C. Paper Flower D. Srakae-bai-dum
     E. Native to Thailand F. Ranong
     G. A climbing, forest shrub with opposite, elliptic leaves; the flowers are small, clustered in heads and dwarfed by six, green, petal-like bracts.

292. A. Verbenaceae B. *Tectona grandis*
     C. Teak D. Sahk
     E. Southeast Asia F. Saraburi
     G. A tall tree which dominates virgin forests in the North and is planted in reforestation projects elsewhere. It has very large, simple, opposite leaves and the flowers are cream-colored and occur in large, loose, terminal panicles.

293. A. Vitaceae B. *Cissus aristata*
     D. Kam-daeng E. Thailand through Indo-Malaysia to New Guinea
     F. Chon Buri
     G. A stout, climbing vine with tendrils and reddish, pubescent stems with broadly-ovate, cordate, deeply-cuspidate-serrate leaves (1). The small flowers are greenish, purple-tinged in leaf-opposed, corymbose cymes and produce large clusters of globose berries (2) which are purple when mature.

294. A. Vitaceae B. *Cissus quadrangularis*
     D. Sam-roi-taw E. Native to Thailand

MIXED FORESTS

    F. Nakhon Ratchasima
    G. A vine of forests and coastal strand with distinctive quadrangular, segmented stems with tendrils at the nodes. The leaves are glabrous, fleshy, broadly-ovate, sharply-toothed and occur at the nodes. The reddish flowers are small and occur in axillary cymes, producing globose, succulent berries. Sometimes cultivated as an ornamental and as a medicinal herb.

295. A. Zingiberaceae   B. *Costus speciosus*
       C. White Costus   D. Uang-din
       E. Indo-Malaysia and Thailand F. Chanthaburi (1,2)
       G. A tall herb growing from a rhizome, with spiraling (1) stem and shiny, oblanceolate, acuminate leaves. The flowers are terminal and borne for a spike of overlapping, green or red bracts. They are white, lobed and have a large lip. There is a *var. argyrophyllus* (2) which is found in more open forests. It is more slender, less-spiralled, grows in larger clumps and has smaller flowers.

296. A. Zingiberaceae   B. *Curcuma spp.*
       C. Wild Ginger   D. Wan Mahamehk
       E. Native to Thailand   F. Prachin Buri (1), Ranong (2)
       G. There are over a dozen similar species of *Curcuma* in Thai forests.These wild gingers are erect, perennial, rhizomatous herbs. They have broad leaves and fleshy inflorescence comprised of spikes of spirally-arranged, showy flowers which sheath the small flowers. Species (1) has pink bracts and yellow flowers and is found in open forests and fields. Species (2) has salmon-colored bracts and white flowers and is found in damp, shady ravine forests.

297. A. Zingiberaceae   B. *Globba spp.*
       D. Kah-ling   E. Native to Thailand
       F. Chanthaburi (1), Chiang Mai (2)
       G. Small, perennial herbs found in ravine forests, they have alternate, lanceolate leaves and drooping, terminal clusters of flowers.*G. leucantha* (1) has white flowers, tipped with purple.*G. obscura* (2) has yellow flowers.

Zingiberaceae

298. A. Zingiberaceae
 B. *Kaempferia pulchra*
 D. Pro-pah
 E. South Thailand and Malaysia
 F. Ranong
 G. A perennial herb with fleshy roots found mostly in damp, shady forests of the Southern Isthmus. It has oval, plicated leaves, splotched with reddish brown. The violet flowers are on short racemes and bloom one or two at a time. The fruit is an ellipsoid capsule about one cm. long.

299. A. Zingiberaceae
 B. *Zingiber spectabile*
 C. Thai Ginger
 D. Wan-prai-dum
 E. Native to Thailand
 F. Chanthaburi
 G. A rhizomatous, forest herb with alternate, oblong-lanceolate leaves. The inflorescence is a bright red spike of overlapping, thick bracts and small, white flowers.

HIGHLAND PINE FOREST

300. A. Balsaminaceae B. *Impatiens masonii*
     E. Native to Thailand F. Nam Nao National Park, Phetchabun
     G. A small herbaceous plant found at high altitudes. It has opposite, linear-lanceolate, slightly-serrate leaves and axillary flowers on long pedicels. The lobed flowers are bright pink and have a long, curling spur. The seeds are within an elongated capsule.

301. A. Compositae B. *Vernonia sp.*
     E. Native to Thailand F. Nam Nao National Park, Phetchabun
     G. A coarse, erect herb with pubescent stems and alternate, oblanceolate, serrate leaves. The purple flower heads occur on terminal spikes and from leaf axils, and produce a globose seed capsule with persistent calyx.

302. A. Convolvulaceae B. *Argyreia splendens*
     D. Krua-khao-luang E. Native to Thailand
     F. Nam Nao National Park, Phetchabun
     G. A trailing or climbing vine with oblong-ovate, shortly-acuminate leaves and short, axillary cymes of funnel-shaped purple blossoms.

303. A. Cycadaceae B. *Cycas pectinata*
     C. Cycad D. Maprao-dao-luang
     E. Native to Thailand F. Nam Nao National Park, Phetchabun
     G. A gymnosperm with an erect, unbranched trunk with a rosette of long, stiff, frond-like leaves and sub-terminal clusters of round, woody fruits on female plants and an erect, center cone on male plants.

Leguminosae

304. A. Dipterocarpaceae B. *Dipterocarpus obtusifolius*
     D. Hieng E. Native to Thailand and Laos
     F. Nam Nao National Park, Phetchabun (1,2)
     G. A medium-height forest tree with large, deeply-plicated, ovate-elliptic leaves (l) that emerge from a red, furled leaf spike. The flowers (2) are showy, 4-lobed and streaked with pink and white.

305. A. Gentianaceae B. *Exacum tetragonum*
     D. Chat-phra-in E. Native to Thailand and Peninsular Malaysia
     F. Nam Nao National Park, Phetchabun
     G. A narrow, erect herb with opposite, lanceolate leaves and panicles of blue-purple and white, 4-lobed flowers in terminal, spike-like clusters.

306. A. Gentianaceae B. *Swertia angustifolia*
     D. Ta-ga-poh E. Native to N. Thailand and the Himalayas
     F. Nam Nao National Park, Phetchabun
     G. An herb with opposite, linear-lanceolate leaves and axillary or terminal, 4-petaled, white flowers with dark flecks.

307. A. Labiatae B. *Leucas lavendulifolia*
     E. India to Thailand F. Nam Nao National Park, Phetchabun
     G. A slender, pubescent herb with opposite, narrowly-ovate, toothed leaves and axillary whorls of tubular, lobed, white flowers.

308. A. Leeaceae B. *Leea acuminata*
     D. Chamaliang-ban E. Native to Thailand
     F. Nam Nao National Park, Phetchabun
     G. A small shrub with pinnate leaves with oblong, serrate leaflets and red joints. The small, white flowers are seated in deep red, corymbose cymes.

309. A. Leguminosae B. *Atylosia volubilis*
     D. Tua-phee E. India to Indochina
     F. Phetchabun

HIGHLAND PINE FOREST

G. A creeping, somewhat scandent vine with trifoliolate leaves and erect racemes of large, yellow pea blossoms. The seed pod is hairy and inflated, the hairs sticky when green. Also seen in other communities.

310. A. Leguminosae B. *Cassia mimosoides*
C. Mimosa-leaved Cassia D. Makam-bia
E. Pantropical F. Nam Nao National Park, Phetchabun
G. An herb with finely-pinnate leaves and axillary, yellow flowers with 5 lobes which are partially closed. The seed pod is about 2 in. long, flat, with 20-25 seeds. Found in other plant communities as well.

311. A. Leguminosae B. *Crotalaria chinensis*
D. Mahing-peh E. Native to Thailand
F. Nam Nao National Park, Phetchabun
G. A hairy, spreading plant with trifoliolate leaves. It is unusual because most species in this genus have yellow flowers; this species has blue-purple blossoms in racemes and produces very fuzzy, inflated pods.

312. A. Leguminosae B. *Crotalaria sessilliflora*
E. Native to Thailand F. Nam Nao National Park, Phetchabun
G. A small, annual, little-branched plant with pubescent, lanceolate leaves and a terminal raceme of yellow or blue pea flowers in a silky calyx.

313. A. Leguminosae B. *Desmodium amethystinum*
E. Native to N. Thailand and China F. Nam Nao National Park, Phetchabun
G. A spreading, mostly prostrate vine with pubescent stems and simple, entire, oblong, slightly-mucronate leaves with short, axillary, stipule-like tufts. The small flowers are deep purple and occur in long, terminal racemes.

314. A. Leguminosae B. *Dunbaria longeracemosa*
D. Krang E. Native to Thailand

300 Impatiens masonii
301 Vernonia sp.
302 Argyreia splendens

303 Cycas pectinata
304.1 Dipterocarpus obtusifolius
304.2 Dipterocarpus obtusifolius

305 Exacum tetragonum
306 Swertia angustifolia
307 Leucas lavendulifolia
308 Leea acuminata

309 Atylosia volubilis
310 Cassia mimosoides
311 Crotalaria chinensis

312 Crotalaria sessilliflora
313 Desmodium amethystinum

314 Dunbaria longeracemosa
315 Eriosema chinense
316 Pueraria phaseoloides

317 Dracaena lourieri
318 Decashistia parviflora
319 Osbeckia chinensis

320 Ardisia pilosa
321 Arundina graminifolia
322.1 Habenaria acuifera

322.2 Habenaria linguella
322.3 Habenaria malintana

322.4 Habenaria rostellifera
323 Christisonia siamensis

324 Buchnera cruciata
325 Triumfetta pilosa

326 Trachydium cambogianum
327 Clerodendron serratum

Melastomaceae

    F. Nam Nao National Park, Phetchabun
    G. A climbing vine with trifoliolate leaves with oblong-elliptic leaflets and long racemes of dark purple and yellow flowers.

315. A. Leguminosae   B. *Eriosema chinense*
        D. Haew-pradu   E. Tropical Asia
        F. Chiang Mai
        G. A small, erect, herbaceous plant with simple, linear-lanceolate leaves and small, yellow, axillary pea blossoms. The tuberous root is reported to be edible.

316. A. Leguminosae   B. *Pueraria phaseoloides*
        D. Tua-sian-pah   E. Indo-Malaya, Thailand
        F. Nam Nao National Park, Phetchabun
        G. A slender, pubescent, twining vine with trifoliate leaves with triangular-ovate leaflets. The flowers are paired, purple and white pea blossoms in tall, erect racemes and produce long, flat, curved pods with about 16 seeds.

317. A. Liliaceae   B. *Dracaena lourieri*
        D. Chan-daeng   E. Native to Thailand
        F. Nam Nao National Park, Phetchabun
        G. A large herb with erect, gray, trunk-like stems and terminal rosette of lanceolate, light green leaves. The flowers are small, white, tubular with 6 lobes in long spikes. Note: The specimens photographed at Nam Nao near the National Park Headquarters were probably transplanted from elsewhere in the park.

318. A. Malvaceae   B. *Decashistia parviflora*
        D. Thong-pan-doon   E. Native to Thailand
        F. Nam Nao National Park, Phetchabun
        G. A few-branched, nearly prostrate, herbaceous shrub with peach-colored blossoms and reddish, pubescent stems with elongated, heart-shaped, blunt leaves.

319. A. Melastomaceae   B. *Osbeckia chinensis*
        D. Ah-noi   E. India to Cambodia

HIGHLAND PINE FOREST

    F. Nam Nao National Park, Phetchabun
    G. A small shrub with pubescent, reddish leaves and terminal, solitary, magenta blossoms with four petals. Found at high altitudes.

320. A. Myrsinaceae   B. *Ardisia pilosa*
       D. Puang-them-hu   E. Native to Thailand
       F. Nam Nao National Park, Phetchabun
       G. A small shrub with reddish stems and oblong-elliptic, glabrous leaves and pendulous umbels of pink, S-lobed flowers which produce red, globose fruits.

321. A. Orchidaceae   B. *Arundina graminifolia*
       D. Yee-toh Penang   E. Native to Thailand
       F. Nam Nao National Park, Phetchabun
       G. A terrestrial with slender, grass-like leaves and erect flower stems. The flowers are showy, violet with a darker purple lip with yellowish throat.

322. A. Orchidaceae   B. *Habenaria spp.*
       E. Native to Thailand   F. Nam Nao National Park, Phetchabun (1,2,3), Ranong (4)
       G. Terrestrials, usually growing from an underground tuber, with slender leaves and upright flower stalks. *H. acuifera* (1) has small, bright yellow flowers. *H. linguella* (2) has white flowers with a long spur. Also with white flowers, but without a spur, is H. malintana (3). *H. rostellifera* (4) also found at the edges of mixed forests, has pink, spider-shaped blossoms.

323. A. Orobanchaceae   B. *Christisonia siamensis*
       D. Dok-din   E. Thailand and Burma
       F. Nam Nao National Park, Phetchabun
       G. A small, nearly stemless parasite consisting of roots and a showy, tubular, purple blossom with a yellow throat. It can be found at the base of grasses in meadows at high elevations.

324. A. Scrophulariaceae   B. *Buchnera cruciata*
       E. Burma, Thailand and Indochina   F. Nam Nao National Park, Phetchabun

Verbenaceae

G. An erect, few-branched herb with opposite, linear-lanceolate leaves and spikes of small, purple blossoms emerging from a conspicuous, angular, herringbone tier of green bracts.

325. A. Tiliaceae B. *Triumfetta pilosa*
C. Hairy Triumfetta D. Oen-galieng
E. Tropical Asia F. Nam Nao National Park, Phetchabun
G. An erect, pubescent herb with ovate, toothed, shortly-acuminate leaves and axillary cymes of yellow flowers with petals closed into tubular shape. The seed is round and covered with long, hairy, hooked spines. Can be found in various plant communities.

326. A. Umbelliferae B. *Trachydium cambogianum*
E. Thailand and Indochina F. Nam Nao National Park, Phetchabun
G. An herb with finely-lobed, fern-like leaves and erect umbels of lacy, white flowers on long peduncles.

327. A. Verbenaceae B. *Clerodendron serratum*
D. Aki-tawan E. Indonesia to Cambodia
F. Nam Nao National Park, Phetchabun
G. A shrub with obovate, shortly-acuminate, serrate leaves and erect, terminal panicles of foliaceous bracts with greenish or bluish-white flowers.

COASTAL STRAND

328.   A. Acanthaceae   B. *Acanthus ebracteatus*
      C. Sea Holly   D. Gnak-plah-maw
      E. Tropical Asia   F. Samut Prakan
      G. An erect, branching herb or undershrub which grows in brackish swamps near the sea, often in association with *Nypa*. It has opposite, shiny leaves which are toothed with stiff spines. The white flowers are borne on an erect spike. A similar species, *A. ilicifolius*, which has blue-violet flowers is also present in Thailand

329.   A. Aizoaceae   B. *Sesuvium portulacastrum*
      C. Sea Purslane   E. Pantropical
      F. Chon Buri
      G. A prostrate, seashore herb found in tidal areas. It has thick, fleshy, linear leaves and reddish stems with roots at the nodes. The flowers are small and pink or violet.

330.   A. Apocynaceae   B. *Cerbera odollam*
      D. Dteen-ped-nam   E. Tropical Asia and Polynesia
      F. Suan Sampran, Rose Garden
      G. A small to medium-sized tree with milky sap and spirally-arranged leaves; most often found in estuaries. It has glossy, oblanceolate leaves and white, 5-lobed flowers. The fruit is green, fibrous, ellipsoid, pendulous and may be solitary or twinned. The sap is poisonous and can cause a rash or blindness.

331.   A. Asclepiadaceae   B. *Tylophora indica*
      E. Native to Thailand   F. Samut Prakan
      G. A woody, twining, shrubby vine in mangroves with glossy, elliptic-ovate, mucronate leaves and short cymes of white, tubular, 5-lobed flowers.

332. A. Boraginaceae B. *Cordia subcordata*
 E. Indo-Malaysia, Thailand and F. Chon Buri
    the Pacific
 G. A small to medium-sized tree with stiff, pale green, broadly-ovate leaves. The flowers are orange and occur mostly at the ends of branches. The fruits are round and occur several together.

333. A. Cactaceae B. *Opuntia vulgaris*
 C. Barbary Fig, Prickly Pear D. Grabong-pet
    Cactus
 E. South Florida, U.S.A. F. Cha-am Beach, Phetchaburi
 G. A succulent plant with multiple tiers of flat, roundish segments which are studded with hard spines. The flowers are orange and occur singly, producing a juicy red fruit which, though edible, is rarely eaten. It is used as a barrier hedge and has escaped to sandy seashores and scrub areas in the Northeast.

334. A. Casuarinaceae B. *Casuarina equisetifolia*
 C. Australian Pine, Ironwood E. Old World Tropics
 F. Rayong
 G. A tall, often narrow tree which is not a true pine, but a flowering plant. The green "needles" are actually linked, scale-like leaves. The male flowers occur at the ends of branches and the female in axils, both small, brown and inconspicuous. The fruit is round, woody and cone-like.

335. A. Chenopodiaceae B. *Suaeda maritima*
 C. Sea Blite E. Java, India to Thailand
 F. Chon Buri
 G. A low shrubby herb found on salt flats (1) near the sea. It has succulent, linear leaves (2) which may be green, red or purple. The flowers are very small, green and occur in clusters.

336. A. Combretaceae B. *Lumnitzera racemosa*
 E. Tropical Asia F. Chon Buri
 G. A woody, mangrove species with glabrous, thick, alternate, oblong-elliptic leaves with acute base and emarginate (slightly-notched)

COSTAL STRAND

apex. The flower are white, 5-petaled and sleeved in an elongated, green calyx. They occur in short, axillary racemes.

337. A. Compositae B. *Pluchea indica*
 C. Indian Marsh Fleabane D. Naht-wua
 E. Tropical Asia F. Thonburi
 G. A many-branched shrub growing 1-2 m high, it has dentate, irregularly-toothed leaves. The pinkish or white flowers occur in a loose, terminal corymb. It is mostly found in coastal wetlands, but is also common in scrub and waste places.

338. A. Compositae B. *Wedelia biflora*
 C. Beach Sunflower D. Pak-krat-talay
 E. Tropical Asia and Pacific F. Samut Prakan
 G. A sprawling shrub found in tidal flats and at the edges of brackish swamps. The leaves are ovate, acuminate, serrate and the flowers are yellow heads.

339. A. Convolvulaceae B. *Calonyction album*
 E. Pantropical F. Koh Samet, Rayong
 G. A scandent, twining vine with orbicular-ovate, acuminate-cordate leaves. The white, tubular flower has a broad, salver-shaped corolla which opens only at night. The fruit is globose and dries into a woody, 4-celled capsule with a persistent 3-part calyx. A close relative, *C. aculeatum* or Moonflower is grown in gardens .

340. A. Convolvulaceae
 B. *Ipomoea pes-caprae ssp. brasiliensis*
 C. Beach Morning Glory D. Pak-boong-talay
 E. Pantropical F. Cha-am Beach, Phetchaburi
 G. A creeping vine found growing in beach sand, it has fleshy, glossy, rounded leaves which are notched at the apex and indented at the base. The flowers are large, rose-purple and trumpet-shaped, and the fruit is a rounded capsule containing fuzzy seeds.

341. A. Convolvulaceae B. *Ipomoea stolonifera*
 E. Pantropical F. Cha-am Beach, Phetchaburi

Guttiferae

    G. A glabrous, trailing vine, rooting at the nodes. The leaves are variably-shaped and two-lobed. The flower is white, pale yellow on the inside, axillary, with funnel-shaped corolla with acute lobes.

342. A. Euphorbiaceae   B. *Glochidion littorale*
        C. Tidal Marsh Glochidion   E. Indo-Malaysia and Thailand
        G. A glabrous shrub found in brackish marshes, the leaves are alternate, entire, elliptic-ovate and slightly-notched at the apex. The greenish flowers are axillary, small and produce a sub-globose, pinkish-white fruit with numerous seeds covered with a red pulp.
        F. Samut Prakan

343. A. Euphorbiaceae   B. *Synostemon bacciformis*
        F. Samut Prakan
        G. A slender, erect, slightly-branched herb with reddish stems and opposite, acute, oblong-linear leaves. The small, white flowers are axillary along the underside of stems and produce a capsule with 6 seeds and a 5-part calyx.

344. A. Goodeniaceae   B. *Scaevola taccada*
        C. Half-flower   F. Koh Samet, Rayong
        G. A many-branched shrub with soft wood and pale green, glossy, oblong leaves, rounded at the apex and narrow at the base. The half-flowers are axillary, white and crepy with 5 lobes.

345. A. Gramineae   B. *Spinifex littoreus*
        C. Spinifex Grass   F. Rayong
        G. A coarse, creeping grass growing in beach sand, rooting at the nodes, with silvery stems and linear-lanceolate leaves. The erect flower heads are comprised of an orbicular cluster of stiff, needle-like awns and small, white flowers.

346. A. Guttiferae   B. *Ochrocarpus siamensis*
        D. Sarapee   F. Suan Thonburi
        G. A medium-sized native tree of the coastal strand, it is also planted in gardens as a handsome shade tree. It has elliptic, smooth, dark green,

COSTAL STRAND

leathery leaves. The flowers are white with numerous yellow stamens. The fruit is green and spherical, about one inch in diameter.

347.
- A. Lauraceae
- B. *Cassytha filiformis*
- E. Pantropical
- F. Koh Samet, Rayong
- G. A leafless, parasitic vine with green or yellow-orange stems that form a tangled mass over host vegetation. The flowers are tiny, white and occur in short spikes and produce a pea-sized berry. This plant is similar in appearance to *Cuscuta*, but not related and has tougher stems.

348.
- A. Lecythidaceae
- B. *Barringtonia racemosa*
- D. Jik-suan
- E. E. Africa, S.E. Asia and the Pacific
- F. Suan Thonburi
- G. A small tree which likes damp coastal swamps, but is also cultivated as an ornamental for its long, hanging strands of pink or white flowers which are comprised of numerous filaments which drop off early in the day. The large leaves, often insect-chewed, are spiraled at the ends of branches.

349.
- A. Leguminosae
- B. *Abrus precatorius*
- C. Crab's Eye, Prayerbead, Coral Bean
- E. Pantropical
- F. Samut Prakan
- G. A branching, woody climber with pinnately-compound leaves. The flowers are pink in axillary racemes and produce pods of glossy, red seeds, each with a black spot. The seeds are deadly poisonous. It can be found in forest, scrub and within the coastal strand.

350.
- A. Leguminosae
- B. *Caesalpinia sappan*
- C. Sappan Wood
- E. India, naturalized in Thailand
- F. Chon Buri
- G. A shrub, sometimes climbing, with prickly, trailing branches and large, bipinnately-compound leaves. The yellow flowers occur in erect, axillary racemes. The pod is short, thick and covered with spines.

328 Acanthus ebracteatus
329 Sesuvium portulacastrum

330 Cerbera odollam
331 Tylophora indica
332 Cordia subcordata
333 Opuntia vulgaris

334 Casuarina equisetifolia
335.1 Suada maritima
335.2 Suada maritima

336 Lumnitzera racemosa
337 Pluchea indica
338 Wedelia biflora

339 Calonyction album
340 Ipomoea pes-caprae ssp. brasiliensis
341 Ipomoea stolonifera

342 Glochidion littorale
343 Synostemon bacciformis
344 Scaevola taccada

345 Spinifex littoreus
346 Ochrocarpus siamensis
347 Cassytha filiformis

348 Barringtonia racemosa
349 Abrus precatorius
350 Caesalpinia sappan

351 Canavalia lineata
352 Derris trifoliata
353 Derris scandens

354 Indigofera hirsuta
355 Tephrosia purpurea
356 Gloriosa superba

357 Pemphis acidula
358 Hibiscus tiliaceus
359 Thespesia populnea

360 Olax scandens
361 Jasminum bifarium
362.1 Nypa fruticans

362.2 Nypa fruticans
362.3 Nypa fruticans
363 Pandanus odoratissimus

364 Portulaca pilosa
365 Guettarda speciosa
366 Citrus microcarpa

367 Sonneratia caseolaris
368 Helicteres hirsuta

369 Clerodendron inerme
370 Premna integrifolia
371 Vitex ovata

Liliaceae

351. A. Leguminosae
 B. *Canavalia lineata*
 E. Pantropical
 F. Koh Samet, Rayong
 G. An herbaceous, prostrate vine with trifoliolate leaves and erect, long-stemmed racemes of pink-purple flowers. Found growing on sandy beaches.

352. A. Leguminosae
 B. *Derris trifoliata*
 D. Tawp-tep-nam
 E. Old World Tropics
 F. Samut Prakan
 G. A creeping or climbing vine-like shrub which is usually found in mangrove or *Nypa* swamps. It has long, trailing branches and compound leaves with mostly 3-5 leaflets. The flowers are small, white and occur in racemes growing from leaf axils. The roots contain rotenone which can be used as a fish poison.

353. A. Leguminosae
 B. *Derris scandens*
 C. Jewel Vine, Tuba Root
 D. Tao-wan-priang
 E. S.E. Asia
 F. Bang Saen, Chon Buri
 G. A woody, forest liana which is also common along the coastal strand. It has compound leaves with oblong-ovate leaflets. The small, white flowers occur on long, drooping racemes.

354. A. Leguminosae
 B. *Indigofera hirsuta*
 E. Pantropical
 F. Ban Phe, Rayong
 G. An erect, shrubby herb with reddish brown stems, all parts pubescent. The leaves are odd-pinnate with 7 oblong-eiliptic leaflets.

355. A. Leguminosae
 B. *Tephrosia purpurea*
 F. Cha-am Beach, Phetchaburi
 G. An undershrub with odd-pinnate leaves with opposite, oblong-elliptic, mucronate leaflets. The flowers are small, purple pea blossoms which occur in a spike-like raceme which is leaf-opposed.

356. A. Liliaceae
 B. *Gloriosa superba*
 C. Gloriosa Lily
 D. Dao-a-dung
 E. Native to Thailand
 F. Bang Saen, Chon Buri
 G. A climbing herb with tendrils at the ends of lanceolate leaves. It grows near the sea in sandy soils of coastal strand forests and is

COSTAL STRAND

cultivated for its showy flowers. The flowers are comprised of six, upturned, wavy, flame-like petals which are yellow and tipped in red. The long stamens radiate from the center of the nodding blossoms.

357.   A. Lythraceae   B. *Pemphis acidula*
         E. E. Africa, Pacific and Tropical   F. Koh Samet, Rayong
             Asia
         G. A shrub or small tree with gnarled grey trunk and hardwood branches. It has small, oblong-elliptic, opposite, grey-green leaves. The small, white flower has 6 crepy petals and is mostly enclosed in a reddish calyx.

358.   A. Malvaceae   B. *Hibiscus tiliaceus*
         C. Hibiscus Tree   D. Paw-talay
         E. Pantropical   F. Samut Prakan
         G. A small tree with low, spreading branches and smooth, fibrous bark. It has large, heart-shaped leaves which are covered with soft hairs. The flowers are bright yellow in the morning, turning reddish, folding and dropping off as the day passes. It is usually found in coastal swamps, but can also grow on higher land.

359.   A. Malvaceae   B. *Thespesia populnea*
         E. Paleotropical   F. Chon Buri
         G. A medium-sized tree with dense foliage of heart-shaped leaves with long, tapering apex, long petioles and conspicuous light yellow leaf veins. The hibiscus-like flowers are yellow in the morning, turning reddish and dropping off by evening.

360.   A. Olacaceae   B. *Olax scandens*
         E. Native to Thailand   F. Cha-am Beach, Phetchaburi
         G. A climbing shrub with ovate-elliptic leaves and subterminal racemes of small, white flowers. The fruit is an orange, ovoid capsule nearly covered by a fleshy calyx.

361.   A. Oleaceae   B. *Jasminum bifarium*
         C. Jasmine   E. Thailand and Malaysia
         F. Chon Buri

Rubiaceae

G. A somewhat scandent, glabrous shrub with trailing branches and ovate leaves. The white flowers have 6-7 lobes and are terminal or axillary. The fruit is a fleshy, globose berry, black when ripe. About two dozen species of native and ornamental jasmines have been recorded in Thailand.

362. A. Palmae   B. *Nypa fruticans*
     C. Nipa   D. Jahk
     E. Southeast Asia   F. Samut Prakan (1,2); Suan Sampran (3)
     G. A palm usually found in pure stands at estuaries or inland of the brackish mangrove zone. It has large, pinnate leaves (1) and a submerged trunk. There are male and female flowers together, the male in thick spikes, the female in a large head (2). The flowers are yellow-orange-brown and produce a basketball-sized brown fruit (3). The phalanges of the fruit have a jelly-like center which is edible. The leaves are valued for thatch.

363. A. Pandanaceae   B. *Pandanus odoratissimus*
     C. Pandanus, Screwpine   E. Tropical Asia
     F. Koh Samet, Rayong
     G. A small tree with forked trunks and prop roots. The leaves, clustered terminally, are long, linear coming to a narrow point and armed with sharp teeth along the margins and midrib. The male inflorescence is pendulous, comprised of white, lanceolate bracts. The fruit is large, ellipsoid and comprised of numerous phalanges which are reddish-yellow when ripe. The leaves are useful for weaving mats.

364. A. Portulaceae   B. *Portulaca pilosa*
     E. Tropical America, naturalized in Thailand   F. Chon Buri
     G. A somewhat prostrate, succulent herb found growing on rocks along rocky coastline. It has small, short, linear leaves and small, reddish purple blossoms.

365. A. Rubiaceae   B. *Guettarda speciosa*
     E. Tropical Asia and Pacific   F. Koh Samet, Rayong

COSTAL STRAND

    G. A small tree with large, opposite, obovate leaves; pubescent on the lower surface. The tubular, white flowers occur in short cymes. The fruit is globose and mostly comprised of a woody endocarp.

366. A. Rutaceae   B. *Citrus microcarpa*
    C. Musk Lime   E. Tropical Asia
    F. Cha-am Beach, Phetchaburi
    G. A small, spreading shrub which can reach a height of about 12 feet. It has thorny branches and slightly shiny, simple leaves with slightly blunt apex with a small notch. The flowers are small, white and have five petals. The fruit is small, round and green, becoming slightly yellowish when ripe and is both sour and seedy. The juice is edible. Some botanists consider this species to be a Malay variation (*var. limau kasturi*) of *C. medica*. Found growing in sandy soil of coastal strand scrub.

367. A. Sonneratiaceae   B. *Sonneratia cascolaris*
    C. Cork Tree   D. Lampoo
    E. India to S.E. Asia   F. Suan Sampran, Rose Garden
    G. A fairly tall tree with simple, pointed, elliptic, opposite leaves. The flowers are a folded bundle of white, filament stamens with a long, protruding pistil which remains attached to the reddish-tinged calyx tube after the flower opens in the early morning and quickly sheds its stamens which litter the ground beneath the tree. It is found near estuaries and mangrove swamps and, like mangroves, it has erect, exposed roots around the base, called pneumatophores, which gather oxygen in areas of tidal inundation.

368. A. Sterculiaceae   B. *Helicteres hirsuta*
    E. Native to Thailand   F. Cha-am Beach, Phetchaburi
    G. An erect shrub found in coastal strand scrub, with oblong-ovate, cordate-acuminate. toothed leaves. The small, violet flowers bloom solitary in short. axillary cymes and produce a shaggy beaked fruit.

369. A. Verbenaceae   B. *Clerodendron inerme*
    E. Indo-Malaysia, Thailand and Pacific   F. Thonburi

Verbenaceae

G. A trailing shrub with woody stems and glossy. opposite leaves. The white, tubular flowers have long. pink stamens. The plant is often used medicinally and is found near the shore and along estuaries.

370. A. Verbenaceae B. *Premna integrifolia*
 E. Tropical Asia and Pacific F. Chon Buri
 G. A scandent. glabrous shrub found near the seashore, it has glossy ovate. opposite. acute-acuminate leaves with rounded base. The small. greenish-white flowers are clustered in large, flat cymes and produce black berries.

371. A. Verbenaceae B. *Vitex ovata*
 E. Coastal Asia F. Cha-am Beach, Phetchaburi
 G. A mostly prostrate shrub which grows in beach sand. The violet flowers are similar to those of *V. trifolia* depicted in the Garden section. This species has simple, ovate leaves and long, prostrate runners radiating from a center cluster of ascending branches.

FRESHWATER MARSH

372. A. Acanthaceae B. *Andrographis paniculata*
     E. India, Malaysia, Thailand F. Trat
     G. An herb with glabrous, lanceolate, acuminate leaves and axillary or terminal, paniculate racemes of small, white flowers. Sometimes cultivated for medicinal use.

373. A. Amaranthaceae B. *Alternanthera philoxeroides*
     C. Alligator Weed E. Native to Thailand
     F. Samut Prakan
     G. A marsh plant, rooting at the nodes with hollow triangular stems and opposite, linear-lanceolate leaves. The white flower heads are on long, axillary peduncles.

374. A. Amaranthaceae B. *Alternanthera sessilis*
     C. Alligator Weed D. Pak-ped-nam
     E. Old World Tropics F. Bangkok
     G. A spreading, prostrate, branched herb found in wet places. The stems are pubescent; the leaves are opposite, elliptic-lanceolate; the small, white flowers are in axillary heads.

375. A. Araceae B. *Pistia stratiotes*
     C. Water Lettuce D. Jauk
     E. Pantropical F. Si Sa Ket
     G. A floating herb with light green leaves in a rosette. It does not flower, but sends out runners to reproduce. Sometimes grown as an ornamental in garden ponds.

376. A. Butomaceae B. *Limnocharis flava*
     C. Yellow Burhead D. Talapat Rusee
     E. Tropical America F. Chiang Mai

Leguminosae

G. An erect. aquatic herb with thick, trigonous petioles and thick. ovate, rounded leaves. The umbelliform flowers are yellow. It is grown in ponds as an ornamental and has escaped to marshes and irrigation ditches.

377. A. Commelinaceae B. *Commelina spp.*
    C. Spiderwort, Dayflower D. Pak-prap
    E. Tropical Asia F. Chiang Mai (I); Bangkok (2)
    G. Prostrate. branched. spreading herbs, somewhat glabrous, with roots emerging from nodes. The leaves are lanceolate and sheathe the stem. The blue flowers emerge from a leaf-like spathe. *C. benghalensis* (l) has small flowers and a large spathe which exudes a mucilaginous fluid when squeezed which is sometimes used in the preparation of traditional medicines. The entire plant is commonly collected and chopped for use as pig feed. Also common in moist habitats is *C. nudiflora* (2) which has narrower leaves and larger flower than the former.

378. A. Convolvulaceae B. *Ipomoea aquatica*
    C. Water Morning Glory D. Boong
    E. S.E. Asia F. Bangkok
    G. A perennial, glabrous vine with hollow stems; the leaves are acute, oblong-ovate and slightly sinuate. The flowers are white or violet and bell-shaped. Found in marshes and cultivated in floating beds for its edible leaf tips.

379. A. Leguminosae B. *Aeschynomene indica*
    C. Sola Plant E. Old World Tropics
    F. Samut Prakan
    G. An erect, branched, annual herb found in wet fields and ditches; it has sensitive, pinnately-compound leaves with small leaflets. The small, light yellow pea flowers have a toothed lip and occur in few-flowered racemes. The seed pod is jointed.

380. A. Leguminosae B. *Sesbania roxburghii*
    D. Sa-no E. S.E. Asia
    F. Bangkok

FRESHWATER MARSH

G. A tall, shrubby herb with a woody base; the leaves are pinnately compound. Bright yellow pea flowers hang in axillary racemes. The flowers are edible.

381.
A. Myrtaceae
B. *Melaleuca leucadendron*
C. Paper Bark
E. Australia, Malaysia and Thailand
F. Chanthaburi
G. A small-medium height tree which can be found in nearly pure stands in low, wet floodplain areas. It may have been early naturalized. The leaves yield and oil known as cajaput. The leaves are oblong-lanceolate with acute apex and base. The flowers are comprised of white filaments and are bottlebrush-like.

382.
A. Nymphaeaceae
B. *Nelumbo nucifera*
C. Sacred Lotus
D. Bua
E. India
F. Ayutthaya
G. A large, perennial, aquatic herb with creeping, underground rhizomes. The leaves are light green and orbicular. The showy flowers are pink or white. The seed pod is slightly woody and has wide openings to allow the seeds to fall out. The seeds are edible.

383.
A. Nymphaeaceae
B. *Nymphaea lotus*
C. Water Lily
D. Bua-sai
E. Pantropical, probably naturalized in Thailand
F. Bangkok (1,2)
G. Aquatic, perennial herbs with stout rootstocks; the leaves are arranged in a spiral and are floating, peltate, deeply cordate and have sinuate margins. The flowers of N. lotus (l) are white or light pink. Those of var. pubescens (2) are dark red. This species has naturalized and the stems are collected as a vegetable. Other Nymphaea species are cultivated in garden ponds.

384.
A. Onagraceae
B. *Jussiaea linifolia*
C. Narrow-leaved Willow Herb
E. Tropical America
F. Samut Prakan

372 Andrographis paniculata
373 Alternanthera philoxeroides

374 Alternanthera sessilis
375 Pistia stratiotes
376 Limnocharis flava

377.1 Commelina benghalensis
377.2 Commelina nudiflora
378 Ipomoea aquatica
379 Aeschynomene indica

380 Sesbania roxburghii
381 Malaleuca leucadendron

382 Nelumbo nucifera

383.1 Nymphaea lotus
383.2 Nymphaea lotus var. pubescens

384 Jussiaea linifolia
385 Ludwigia adscendens
386 Ludwigia octovalvis
387 Polygonum pulchrum

389 Monochoria hastata

388 Eichhornia crassipes
390 Typha angustifolia

391 Xyris pauciflora

382 Nelumbo nucifera

Pontederiaceae

G. A naturalized annual with angled stems and lanceolate. leaves. The solitary yellow flowers have four petals and four green sepals which are as long as the petals and can be seen between the petals.

385. A. Onagraceae B. *Ludwigia adscendens*
 C. Creeping Water Primrose D. Pang-poey-nam
 E. Tropical Asia F. Bangkok
 G. An erect, branched, glabrous herb found in wet places. It roots at the nodes and the leaves are oblong-elliptic. The cream-colored to light yellow flowers occur singly at upper leaf axils.

386. A. Onagraceae B. *Ludwigia octovalvis*
 C. Water Primrose D. Tian-nam
 E. Old World Tropics F. Chanthaburi
 G. An erect, annual marsh herb, often persisting and becoming woody at the base. It is branching, often with purplish stems. The leaves are lanceolate and it has yellow, 4-petaled flowers growing from upper leaf axils.

387. A. Polygonaceae B. *Polygonum pulchrum*
 C. Watersmart Weed D. Pai-nam
 E. Tropical Asia and Africa F. Ayutthaya
 G. An aquatic weed found in marshes and roadside ditches; it has hollow stems, roots at the nodes and reaches a length of 2 m., much of which is submerged. The leaves are very pubescent, lanceolate, acuminate and have basal sheaths encircling the stem. The small, white flowers occur in panicled racemes which are erect or nodding.

388. A. Pontederiaceae B. *Eichhornia crassipes*
 C. Water Hyacinth D. Pak-top-chawa
 E. Tropical America F. Bangkok
 G. A floating herb which sometimes roots into mud; it has a short stem with broad, rounded leaves and swollen, spongy petioles. The pale violet flowers cluster on an erect spike.This plant has naturalized to the extent that it is now considered to he the most noxious aquatic weed pest in S.E. Asia. clogging waterways. It is sometimes collected for use as pig food.

FRESHWATER MARSH

389. A. Pontederiaceae   B. *Monochoria hastata*
      C. Hastate-leaved Pondweed   D. Pak-top Thai
      E. S.E. Asia   F. Bangkok
      G. An erect, glabrous herb with thick. broadly ovate leaves. The flowers emerge from the sheathed, thickly cylindrical petiole near the base of the leaf.

390. A. Typhaceae   B. *Typha angustitolia*
      C. Cattail   D. Yah-toop
      E. Pantropical   F. Bangkok
      G. A perennial. erect marsh herb with linear leaves which sheath at the base of the stem. The flowers are small and crowded into a long brown. cylindrical spike.

391. A. Xyridaceae   B. *Xyris pauciflora*
      E. Tropical Asia
      G. An erect, grass-like plant found in ricefields, wet grasslands and roadside ditches. The leaves are rigid, linear and acuminate at the apex. the small yellow flowers are borne on an ovoid, brown head of overlapping bracts which is terminal on a long, erect spike. the photograph was taken of specimens sold in the market. The author has seen Xyris in the wild in Phuket and along the highway between Krabi and Trang in Southern Thailand.

FIELD

392. A. Acanthaceae B. *Asystasia gangetica*
     C. Indian Asystasia E. India and Sri Lanka
     F. Cha-am Beach, Phetchaburi
     G. An erect herb growing to a height of about 2 ft. The bell-shaped flowers are light violet with a light yellow throat or completely light yellow and occur in spikes. Sometimes planted as an ornamental, it has also naturalized along roadsides.

393. A. Acanthaceae B. *Ruellia tuberosa*
     C. Popping Pod E. S.E. Asia
     F. Bangkok
     G. A low, perennial herb with opposite leaves. The tubular, bell-shaped flower is purple and the fruit is a cylindrical capsule.

394. A. Amaranthaceae B. *Achyranthes aspera*
     C. Chaff Flower E. Pantropical
     F. Bang-pa-in, Ayutthaya
     G. A coarse, branched, annual herb with oblong-obovate leaves. The flowers are green and occur in erect, silvery spikes. The plant is reported to have medicinal qualities.

395. A. Amaranthaceae B. *Gomphrena celosiodes*
     C. Globe Amaranth D. Ban-mai-ru-roi-pah
     E. Tropical America F. Bangkok
     G. A low, creeping herb with hairy stems and simple, opposite leaves. The white flowers are densely crowded in erect, terminal, globose heads.

396. A. Apocynaceae B. *Catharanthus roseus*
     C. Periwinkle D. Phaeng-phuai-farang
     E. Madagascar F. Rayong

FIELD

    G. A garden plant which has naturalized in waste places; an erect, branching herb with milky sap and obovate leaves. The axillary flowers are 5-lobed and purple-pink or white.

397. A. Araceae   B. *Typhonium trilobatum*
       C. Three-lobed Typhonium   D. Utaphit
       E. Native to Thailand   F. Bang-pa-in, Ayutthaya
       G. A fleshy, annual herb with grows from an underground tuber. It is found in shady, waste places. The leaves are arrow-shaped and trilobed. The flower is a reddish spike with a large, brown spathe and has an unpleasant odor.

398. A. Asclepiadaceae   B. *Calotropis gigantea*
       C. Giant Indian Milkweed,   D. Ruk
          Crown Flower, Ivory Plant
       E. India to S.E. Asia   F. Nakhon Ratchasima
       G. A large shrub or small tree found growing in dry, sandy soils. It has a milky sap and the broad, opposite leaves have a silvery, powdery appearance. The flowers are purple or white and are often used in flower garlands.

399. A. Asclepiadaceae   B. *Oxystelma esculentum*
       D. Chamuuk-plah-lai-dong   E. Native to Thailand
       F. Chon Buri
       G. A vine with milky sap found in scrub and waste places with opposite, lanceolate leaves, dark green with conspicuous light green veins on top and light green on the underside. The flowers are bell-shaped with 5 lobes, white with reddish streaks inside.

400. A. Boraginaceae   B. *Heliotropium indicum*
       C. Wild Clary, Indian Turnsole   D. Yah-nguang-chang
       E. Pantropical   F. Chon Buri
       G. An erect, annual, branched herb found in waste places. The leaves are opposite or alternate, hairy and ovate. The flowers are small, pale violet to white and occur on one side of a coiled, terminal spike. The fruit is small and comprised of two nutlets.

Compositae

401. A. Capparidaceae B. *Cleome viscosa*
 C. Yellow Cleome D. Pak-sian-pee
 E. Pantropical F. Samut Prakan
 G. An erect, branched herb with 3-9-foliate leaves. The yellow flowers occur in terminal racemes. The stems and seed pods are pubescent.

402. A. Capparidaceae B. *Gynandropsis pentaphylla*
 C. Spider Flower, Bastard Mustard D. Pak-sian
 E. Pantropical F. Cha-am Beach, Phetchaburi
 G. Also listed as *Cleome gynandra*, this is an erect, slightly pubescent herb with leaves divided into 5 leaflets. The flowers occur on an elongated raceme and are white or purplish with extremely long ovaries. The cylindrical fruit occur on an extremely long petiole, thus the herb has a very unusual structure. It is edible and usually pickled

403. A. Compositae B. *Eclipta prostrata*
 C. False Daisy D. Kameng
 E. Pantropical F. Bangkok
 G. A small, branching, annual herb with narrow leaves and small, white flower heads. Also listed as *E. alba*.

404. A. Compositae B. *Elephantopus scaber*
 C. Elephant's Foot D. Yah-fai-nok-khum
 E. Tropical Asia F. Ranong
 G. A small, hairy herb with leaves in a rosette, it grows from creeping rhizomes. The purple or white filament-like flower heads are enclosed in three green bracts.

405. A. Compositae B. *Erechthites valerianifolia*
 C. Malayan Groundsel, Fireweed E. Tropical America
 F. Ranong
 G. A naturalized herb in waste places, it has alternate, deeply-toothed leaves and often nodding, pinkish-orange flower heads.

406. A. Compositae B. *Eupatorium odoratum*
 C. Jack-in-the-bush D. Yah-dok-khao, Yah-farangset

FIELD

      E. Central America   F. Ranong
      G. A large, shrubby, branching herb with triangular leaves with serrate margins. The white flowers are comprised of thread-like flowerlets in heads and are bunched in corymbs.

407. A. Compositae   B. *Spilanthes acmella*
     C. Para Cress, Ear-stud Flower   D. Pakat-hua-waen
     E. Tropical America   F. Ranong
     G. An erect herb found along roadsides and in scrub, it has simple, pubescent, opposite leaves and yellow flowers in dense, terminal and axillary heads.

408. A. Compositae   B. *Tridax procumbens*
     C. Coat Buttons   D. Dteen-tukkae
     E. Tropical America   F. Bangkok
     G. An herb commonly found in waste places, it has fuzzy stems and coarsely-toothed leaves. The white and yellow flower is terminal and occurs on a long stem.

409. A. Compositae   B. *Vernonia cinerea*
     C. Little Ironweed   D. Yah-sam-wan
     E. Tropical Asia   F. Bangkok
     G. An erect, annual herb, the lower leaves are petiolate and larger than the upper leaves which are narrow and sessile. The small, purple flower heads occur in terminal corymbs. About a dozen different species of *Vernonia* are found in Thailand.

410. A. Convolvulaceae   B. Cuscuta chinensis
     C. Dodder   D. Foi-mai
     E. China   F. Samut Prakan
     G. A parasitic vine with yellow-orange, thread-like stems. The leaves are reduced to minute scales. The flowers are small and creamy white. The host is most often *Wedelia sp.*; thus, it can also be seen in the coastal strand as well as roadsides.

411. A. Convolvulaceae   B. Hewittia sublobata
     D. Chingcho-lek   E. Tropical Asia
     F. Chon Buri

Cucurbitaceae

    G. A twining vine with heart-shaped, acuminate, cordate leaves. The flowers are white with a pale yellow star formed by folds in the corolla. The throat is reddish and the seeds are within a globose capsule.

412. A. Convolvulaceae   B. *Ipomoea indica*
    C. Blue Morning Glory   E. Pantropical
    F. Bangkok
    G. A twining vine with ovate-orbicular, acuminate-cordate leaves which are three-lobed on juvenile plants. The broad, light blue corolla is extremely fragile and folds up soon after the sun rises.

413. A. Convolvulaceae   B. *Ipomoea digitata*
    C. Railway Creeper   D. Pak-boong-rua
    E. Indonesia   F. Chanthaburi
    G. A creeping, sometimes climbing vine with attractive purple flowers with dark throats. The leaves are deeply-lobed.

414. A. Convolvulaceae   B. *Ipomoea obscura*
    D. Sa-uek   E. Tropical Asia
    F. Bangkok
    G. A twining vine often seen on fences, it has reddish stems and heart-shaped leaves. The corolla is light pink with a dark center.

415. A. Convolvulaceae   B. *Merremia gemella*
    C. Yellow Morning Glory   E. Tropical Asia and Africa
    F. Phatthalung
    G. A climbing, creeping vine found along roadsides and in waste places, it has bright yellow flowers and heart-shaped leaves.

416. A. Convolvulaceae   B. *Merremia hirta*
    E. Tropical Asia   F. Rayong
    G. A slender, twining vine reaching a length of 1-2 m. with narrow, oblanceolate leaves. The pale yellow flowers are axillary, solitary or paired. The seed is within a papery, ovoid capsule.

417. A. Cucurbitaceae   B. *Coccinea indica*
    D. Tam-lueng   E. Native to Thailand

FIELD

    F. Nakhon Ratchasima (1,2)
    G. A climbing vine with tendrils and fleshy, lobed leaves. The flowers (1) are solitary, white and 5-lobed. The fruits (2) turn bright red when ripe. The leaf tips are favored as a cooked vegetable and the plant is often cultivated or encouraged for this purpose.

418. A. Cucurbitaceae   B. *Momordica charantia*
        C. Bitter Melon, Bitter Cucumber, Balsam Pear
        D. Ma-hoi   E. S.E. Asia
        F. Samut Prakan
        G. A climbing vine with tendrils and deeply-lobed leaves. It has yellow, solitary, axillary flowers and a warty, cylindrical fruit which is yellow or orange when ripe and contains seeds covered with a bright orange pulp. The fruit is edible, but bitter and is often cultivated in hybrid form.

419. A. Cucurbitaceae   B. *Trichosanthes cucumerina*
        D. Buap-khom   E. Tropical Asia
        F. Phitsanulok
        G. A scandent, herbaceous vine with 3-5-lobed leaves and long-peduncled, axillary flowers. The flowers are usually solitary and are large, yellow with 5 crepy lobes. The fruit is orange-red and ellipsoid when mature.

420. A. Euphorbiaceae   B. *Acalypha indica*
        C. Common Acalypha   D. Tamyae Maeo
        E. Tropical Asia and Africa   F. Ayutthaya
        G. An erect, branched herb with long-petioled, ovate serrate leaves. The small, green, male flowers occur on axillary spikes at the tip and the female flowers are at the base and enclosed in large, green, cup-shaped bracts.

421. A. Euphorbiaceae   B. *Euphorbia cyathopora*
        C. Dwarf Poinsettia   E. Tropical America
        F. Chon Buri
        G. A small, erect herb with milky sap and alternate, lobed leaves, the uppermost with irregular red blotches near the base. The flowers are small, yellowish with 1-2 small glands.

392 Asystasia gangetica
393 Ruellia tuberosa
394 Achyranthes aspera

395 Gomphrena celosiodes
396 Catharanthus roseus
397 Typhonium trilobatum

398 Calotropis gigantea
399 Oxystelma esculentum

400 Heliotropium indicum
401 Cleome viscosa
402 Gynandropsis pentaphylla
403 Eclipta prostrata

404 Elephantopus scaber
405 Erechthites valerianifolia
406 Eupatorium odoratum

407 Spilanthes acmella
408 Tridax procumbens
409 Vernonia cinerea

410 Cuscuta chinensis
413 Ipomoea digitata

411 Hewittia sublobata
412 Ipomoea indica
414 Ipomoea obscura

415 Merremia gemella
416 Merremia hirta

417.1 Coccinea indica
417.2 Coccinea indica
418 Momordica charantia

419 Trichosanthes cucumerina

420 Acalypha indica
421 Euphorbia cyathopora

422 Euphorbia geniculata
423 Jatropha gossypifolia
426 Anisomeles indica

424 Phyllanthus reticulatus
425 Coix lachryma-jobi

427 Hyptis capitata
428 Hyptis suaveolens
429 Leonotis nepetifolia

430 Leucas aspera
431 Ocimum sanctum
432 Alysicarpus nummarifollus

433 Canavalia rosea
434 Cassia occidentalis

435 Cassia tora
436 Centrosema pubescens
437 Clitoria ternatea

438 Crotalaria mucronata
439 Desmanthus virgatus
440 Desmodium captitatum

441 Desmodium triquetrum
442 Dolichos lablab
443 Mimosa pigra
444 Mimosa pudica

445 Mucuna pruriens
446 Phaseolus calcaratus

447 Phaseolus lathyroides
448 Uraria crinata
449 Abelmoschus moschatus

450 Abutilon indicum
451 Hibiscus sabdariffa
452 Malachra captitata

453 Sida acuta
454 Sida retusa
455 Urena lobata

456.1 Nepenthes gracilis
456.2 Nepenthes gracilis

456.3 Nepenthes thorlii
457.1 Boerhaavia diffusa
457.2 Boerhaavia repanda
458 Biophytum sensitivum

459 Oxalis corniculata
460 Passiflora foetida
461 Rivinia humilis

462 Salomonia ciliata
463 Paederia tomentosa
464 Cardiospermum halicacabum

465 Lindenbergia philippensis
466 Striga asiatica
467 Datura metel
468 Physalis angulata

469 Solanum torvum
470 Solanum trilobatum
471 Melochia corchorifolia

472 Waltheria americana
473.1 Corchorus acutangulus
473.2 Corchorus olitorius

474 Lantana camara
475 Lippia nodiflora
476 Stachytarpheta indica
477 Tribulus terrestris

Labiatae

422. A. Euphorbiaceae B. *Euphorbia geniculata*
 C. Common Spurge E. Mexico, Texas and the W. Indies
 G. A naturalized herb with milky sap and hollow stems; the leaves are oblanceolate with narrow apex, some with irregularly-indented margins near the base. The light green flowers occur in terminal clusters. Commonly found in waste places.

423. A. Euphorbiaceae B. *Jatropha gossypifolia*
 D. Saboo-daeng E. Brazil
 F. Thonburi
 G. A shrub with 3-5-lobed, glossy leaves and hairy leaf stems. The young leaves are purplish. The flowers are small, dark red and produce green, capsule-like fruit. The shrub is sometimes planted as an ornamental, but has become a noxious pest, having naturalized in open fields where it is poisonous to grazing cattle.

424. A. Euphorbiaceae B. *Phyllanthus reticulatus*
 E. Tropical Asia and Africa F. Chanthaburi
 G. An erect, somewhat scandent shrub with pendulous branches; the leaves are elliptic-oblong. The small, green flowers are tinged with purple and are axillary, solitary or few on slender pedicels, and produce a fleshy, black, globose berry.

425. A. Gramineae B. *Coix lachryma-jobi*
 C. Job's Tears D. Duei
 E. Tropical Asia F. Bangkok
 G. A tall, branching grass; the ovoid beads enclose female flowers and male flowers project from the tips of beads and drop off at maturity. The hard beads are often strung into necklaces.

426. A. Labiatae B. *Anisomeles indica*
 E. Tropical Asia F. Phangnga
 G. An erect, coarse herb with ovate, serrate leaves and whorled, spike-like racemes of purple flowers which are conspicuous, ovoid and have hairy, acuminately-toothed calyx.

FIELD

427. A. Labiatae B. *Hyptis capitata*
      C. Buttonweed E. Tropical America
      F. Trang
      G. A coarse herb with square stems and serrate leaves. The small, white flowers occur in round heads.

428. A. Labiatae B. *Hyptis suaveolens*
      C. Bush Teabush D. Karah
      E. Tropical America F. Rayong
      G. A coarse, erect, branched, hairy, annual herb with very aromatic, ovate, serrulate, opposite leaves and 4-sided stems. The small, violet flowers are axillary and occur 3-4 together.

429. A. Labiatae B. *Leonotis nepetifolia*
      D. Chat-phra-in E. Tropical Africa
      F. Ban Phe, Rayong
      G. An erect, coarse herb growing to a height of 2 m. with 4-sided, grooved stems. The lower leaves are ovate, toothed; the upper floral leaves are opposite, lanceolate, toothed with an elongated, narrow base. The tubular, red-orange flowers emerge from an axillary, orbicular head of stiff, spiny, green calyxes. Found in waste places.

430. A. Labiatae B. *Leucas aspera*
      D. Yah-hua-toh E. Tropical Asia
      F. Chanthaburi
      G. An erect, pubescent herb with white flowers occurring in dense, terminal heads. The plant's lower leaves are lobed and larger than the leaflets that grow from the upper branches. The leaves have serrate margins.

431. A. Labiatae B. *Ocimum sanctum*
      C. Sacred Basil D. Maeng-luk
      E. Tropical Asia F. Cha-am Beach, Phetchaburi
      G. Several species of basil are cultivated in Thailand. This species is also cultivated, but is commonly found in fields as well and was

Leguminosae

probably early naturalized. It is a branched herb with very small, white flowers emerging from calyxes clustered along an erect, terminal raceme. Used as a spice in food preparation.

432. A. Leguminosae B. *Alysicarpus nummularifollus*
      C. Buffalo Clover E. Old World Tropics
      F. Chon Buri
      G. A spreading, branched herb with leaves varying from elliptic to oblanceolate, with small stipules at the base. The small, dark pink pea blossoms occur in erect, leaf-opposed racemes.

433. A. Leguminosae B. *Canavalia rosea*
      C. Greater Seabean D. Tua-kla
      E. Pantropical F. Nakhon Pathom
      G. A trailing vine found on sandy seashores and in open fields. It has shortly-acuminate, trifoliate leaves. The flowers are rose pink, notched at the tip and occur several together on a spike.

434. A. Leguminosae B. *Cassia occidentalis*
      C. Coffee Senna D. Khee-lek-phee
      E. Tropical America F. Rayong
      G. This species is an introduced roadside herb or small, branching shrub. It has compound leaves and yellow flowers. The seed pods are long and slender. The leaf tips are edible.

435. A. Leguminosae B. *Cassia tora*
      E. Tropical America F. Rayong
      G. An erect, glabrous annual, less than 1 m. high with pinnate, obovate leaves, 6 leaflets and axillary, yellow flowers and a long, thin seed pod. Found in waste places.

436. A. Leguminosae B. *Centrosema pubescens*
      D. Tua-lai E. Tropical America
      F. Chanthaburi
      G. A climbing, creeping vine with trifoliolate leaves and pubescent stems. The flower is purple and pealike.

FIELD

437. A. Leguminosae B. *Clitoria ternatea*
     C. Butterfly Pea, Blue Pea D. Uang-chan
     E. Tropical America F. Bangkok
     G. A climbing vine with dense foliage of compound leaves and bright, violet-blue flowers with white centers. The seeds are within a flat pod about 2 in. long. A double-flowered variety is seen in gardens. Rarely seen is the all-white-blossomed *C. alba*.

438. A. Leguminosae B. *Crotalaria mucronata*
     C. Rattlebox E. Old World Tropics
     F. Cha-am Beach, Phetchaburi
     G. An erect, small shrub with trifoliate leaves. The yellow flowers grow in long-spiked racemes and produce inflated seed pods.

439. A. Leguminosae B. *Desmanthus virgatus*
     E. Tropical America F. Bangkok
     G. An erect shrub with green, striated stems and finely-divided, compound leaves. The axillary flower is a loose head of white, filament-like parts and produces a long, slender pod.

440. A. Leguminosae B. *Desmodium capitatum*
     E. Tropical Asia F. Chon Buri
     G. A sprawling plant with trifoliolate leaves with elliptic leaflets and axillary racemes of small, purple pea blossoms.

441. A. Leguminosae B. *Desmodium triquetrum*
     E. Tropical Asia F. Ranong
     G. An erect, branched herb or undershrub with oblanceolate leaves with prominently-winged petioles. The small, purplish-pink flowers occur in terminal and axillary racemes. Can also be found in the highland pine forests.

442. A. Leguminosae B. *Dolichos lablab*
     C. Hyacinth Bean D. Tua-paep
     E. Old World Tropics F. Samut Prakan
     G. A glabrous, twining annual vine with trifoliolate leaves and long, erect racemes of purple-pink or white flowers. The flattened,

Leguminosae

acuminate seed pods are edible. Grows wild in previously-cleared areas and is sometimes cultivated.

443.   A. Leguminosae   B. *Mimosa pigra*
      C. Giant Sensitive Plant   D. Maiyarap
      E. South America   F. Chiang Mai
      G. A shrub growing 1-2 m. high and commonly found on river banks in N. Thailand, but can also be found in waste places. It has prickly stems and finely-compound leaves. The flowers are pink, filament-puff balls and the seeds are fuzzy, yellow-green pods which occur in clusters.

444.   A. Leguminosae   B. *Mimosa pudica*
      C. Sensitive Plant   D. Ra-ngap
      E. Pantropical   F. Bangkok
      G. A mostly prostrate, but occasionally ascending, creeping plant with prickly, hairy stems and finely-compound leaves which fold when touched. The flowers are round heads of pink filaments and the seed pods are flat and hairy.

445.   A. Leguminosae   B. *Mucuna pruriens*
      D. Mamui   E. India-S.E. Asia
      F. Phitsanulok
      G. A twining vine with trifoliolate leaves and purple-black flowers in pendulous racemes. The seed pods are covered with stiff, brown, irritating hairs and should be avoided.

446.   A. Leguminosae   B. *Phaseolus calcaratus*
      E. Tropical Asia   F. Bang-pa-in, Ayutthaya
      G. Also identified as *Vigna umbellata*, this is an herbaceous, slender, pubescent vine with trifoliolate leaves with oblong-ovate, acute or acuminate, shallowly-lobed leaflets. The yellow flowers occur several together at the apex of a long-peduncled raceme.

447.   A. Leguminosae   B. *Phaseolus lathyroides*
      C. Scarlet Bean   D. Tua-phee
      E. Tropical America   F. Bangkok

FIELD

    G. An erect, branched plant, growing about 1 m in height. The leaves are trifoliate and the flowers pea-like, dark crimson and produce a thin pod.

448. A. Leguminosae   B. *Uraria crinita*
       C. Cat's Tail   D. Hahng-kra-rok
       E. Tropical Asia   F. Chanthaburi
       G. A spreading, perennial herb with 1-3-foliate leaves. The light violet flowers are many and occur in a dense, erect raceme.

449. A. Malvaceae   B. *Abelmoschus moschatus*
       C. Musk Mallow   D. Fai-phee
       E. India-S.E. Asia   F. Nakhon Pathom
       G. The wild variety of this species is an erect, hairy herb with lobed leaves, usually with 3-5 lobes, and large, bright yellow flowers with a dark crimson center.

450. A. Malvaceae   B. *Abutilon indicum*
       C. Country Mallow   E. S.E. Asia
       F. Bangkok
       G. A branching plant with broadly-ovate leaves with irregularly-toothed margins. The leaves and stems are lightly-pubescent, giving them a soft appearance. The flower is yellow and the fruits are drum-shaped and comprised of radiating carpels.

451. A. Malvaceae   B. *Hibisus sabdariffa*
       C. Roselle, Red Sorrel   D. Krachiap-daeng
       E. India   F. Ranong
       G. This is a cultivated, annual herb which has naturalized in open fields, particularly in S. Thailand. It has glabrous, lobed leaves and red stems. The flower is pink with a dark center and produces a fleshy, acid, red fruit comprised of a persistent calyx. The fruit is used for jam, flavoring and dye.

452. A. Malvaceae   B. *Malachra capitata*
       E. Tropical America   F. Bangkok
       G. A coarse herb found in waste places, it is recognized by its angular leaves, hairy stems and yellow flowers. The hairs are irritating and should be avoided.

Nyctaginaceae

453. A. Malvaceae B. *Sida acuta*
     E. Pantropical F. Rayong
     G. A small, perennial, shrub-like herb with lanceolate, serrate leaves and axillary, pale yellow flowers. Common in waste places.

454. A. Malvaceae B. *Sida retusa*
     E. Pantropical F. Chon Buri
     G. An erect, branched, pubescent plant, usually less than .5m. in height. The leaves are obovate and toothed. The flowers are pale yellow, axillary and usually solitary.

455. A. Malvaceae B. *Urena lobata*
     C. Bur-fruited Urena D. Khee-khrok
     E. Pantropical F. Rayong
     G. An erect, branched, coarse herb with angled-lobed, ovate leaves which are irregularly-toothed. The flowers are axillary, solitary, and pink. The fruit is globose with five carpels and is covered with short, stiff spines.

456. A. Nepenthaceae B. *Nepenthes gracilis*
     C. Pitcher Plant D. Nam-tao
     E. Native to Thailand and Malaysia F. Ranong (1,2,3)
     G. This species is found in open, sunny areas with sandy soils. The unique, complex leaves have an erect, tubular appendage with a flap to regulate water content (1). Insects drown in the liquid and a digestive fluid decomposes them so the leaves can absorb the nutrients with the water. The flowers are grayish-green and borne on long, erect spikes. As the spike matures, golden brown pods filled with wind-borne, fiber-like seeds develop (2). Less commonly seen is *N. thorlii* (3) which has biwinged pods which are more squat-like than those of the former. This species does not creep or climb like the former, is more compact and prefers more shade.

457. A. Nyctaginaceae B. *Boerhaavia spp.*
     C. Spreading Hogweed E. Tropical Asia to Polynesia
     F. Bangkok (1), Bang-pa-in, Ayutthaya (2)
     G. *B. diffusa* (l) is a spreading, branched herb with glabrous, mostly-ovate leaves. The small, purple flowers occur in long, lax, panicled

cymes. The fruit is glandular. *B. repanda* (2) is a more woody herb with triangular, ovate, senuate leaves and has short, stiff cymes of pink flowers.

458. A. Oxalidaceae   B. *Biophytum sensitivum*
     E. Pantropical   F. Bangkok
     G. An annual, small, unbranched herb with pinnate leaves. The small, yellow flowers are at the ends of peduncles as long as the leaves.

459. A. Oxalidaceae   B. *Oxalis corniculata*
     C. Wood Sorrel   E. Paleotropical
     F. Thonburi
     G. A ground cover herb found in waysides, lawns and as a weed in garden pots. It has trifoliate leaves with heart-shaped leaflets and small, yellow flowers in cymes.

460. A. Passifloraceae   B. *Passiflora foetida*
     C. Love-in-a-mist   D. Ka-thok-rok
     E. Tropical America   F. Ranong
     G. A climbing, creeping, perennial vine with hairy, 3-lobed leaves. The flower is purple and the thin-skinned fruit is round, orange and filled with black seeds. The leaf tips are used as a cooked vegetable.

461. A. Phytolaccaceae   B. *Rivinia humilis*
     C. Coral Berry   D. Prik-farang
     E. Tropical America   F. Bangkok
     G. An herb with woody base and ovate-lanceolate leaves. The small, white flowers are in terminal or axillary racemes and produce small, red berries. The plant contains a skin irritant.

462. A. Polygalaceae   B. *Salomonia ciliata*
     C. Common Salomonia   E. Tropical Asia
     F. Ranong
     G. A small, erect, annual herb with glabrous, heart-shaped leaves with acute apex and rounded, sessile base. The small, purple-pink flowers are borne on slender, terminal spikes. It is found in dry, grassy areas.

Solanaceae

463. A. Rubiaceae B. *Paederia tomentosa*
 E. India to Japan F. Phatthalung
 G. A slender, twining, herbaceous vine with a strong odor when crushed. It has heart-shaped leaves and few-flowered, axillary cymes of small, white flowers with deep purple throats.

464. A. Sapindaceae B. *Cardiospermum halicacabum*
 C. Baloon Vine D. Khok-kra-om
 E. India F. Bang-pa-in, Ayutthaya
 G. A slender, herbaceous, climbing vine with trifoliate, alternate, deeply-lobed and coarsely-toothed leaves. The flowers are very small and white. The fruit is an inflated, 3-sided pod and contains black seeds.

465. A. Scrophulariaceae B. *Lindenbergia philippensis*
 E. Tropical Asia F. Ayutthaya
 G. A small, erect, annual, hairy herb with basal, obovate, toothed leaves and terminal racemes of tubular, yellow flowers. Likes to grow in crevices in old walls in temple ruins.

466. A. Scrophulariaceae B. *Striga asiatica*
 C. Witchweed E. Tropical Asia
 F. Ranong
 G. An erect, slender herb with pubescent, lanceolate leaves and tiny, yellow, lobed flowers. This plant is parasitic, attaching itself to grass roots by means of suckers which penetrate the host plant. Seen in fields and roadsides or within the coastal strand.

467. A. Solanaceae B. *Datura metel*
 C. Angel's Trumpet D. Lampong
 E. Tropical Asia F. Samut Prakan
 G. Probably naturalized, this is an herb or short-lived shrub which is usually seen only in waste places. It has large, ovate leaves with irregular margins. The long, trumpet-shaped flowers are light yellow or white and the fruits are round and studded with short thorns. The plant has a very noxious odor and is poisonous if ingested.

FIELD

468. A. Solanaceae B. *Physalis angulata*
      C. Cape Gooseberry E. Tropical America
      F. Chiang Mai
      G. A smooth-leaved herb with angular stems and ovate, dentate leaves. The flowers are light yellow. The fruit is ellipsoid and contained within an inflated calyx that resembles a Japanese lantern.

469. A. Solanaceae B. *Solanum torvum*
      C. Wild Eggplant D. Mah-kua-puang
      E. Tropical Asia F. Nakhon Ratchasima
      G. A shrubby herb found in waste places and sometimes cultivated for its fruit, it has broad leaves with uneven margins. The flowers are small, white and produce round, green fruits which occur in clusters.

470. A. Solanaceae B. *Solanum trilobatum*
      D. Ma-waeng-khruea F. Cha-am Beach, Phetchaburi
      G. A woody, climbing shrub with lobed leaves and prickly stems. The flowers are purple with bright yellow stamens. The fruit is a bright red berry when ripe.

471. A. Sterculiaceae B. *Melochia corchorifolia*
      E. Pantropical F. Bangkok
      G. An erect, branched herb with oblong-ovate leaves. The small, pale pink flowers are in crowded, axillary or terminal heads. The seed is a depressed-globose capsule.

472. A. Sterculiaceae B. *Waltheria americana*
      E. Tropical America F. Chon Buri
      G. An erect, slightly-branched, pubescent plant with oblong-ovate, rounded, toothed, prominently-nerved leaves. The small, yellow flowers are in axillary clusters.

473. A. Tiliaceae B. *Corchorus spp.*
      E. Tropical Asia F. Suan Thonburi (1), Nakhon Chaisri, Nakhon Pathom (2)
      G. *C. acutangulus* (1) is a branching, annual field herb less than 1 m. in height. It has reddish stems and oblong-ovate leaves which are

Zygophyllaceae

finely-toothed. The flowers are axillary, yellow and have a base of 5 leafy bracts which project beyond the petals. A similar species, naturalized from India, *C. olitorius* (2) has ovate-lanceolate leaves which are winged at the base.

474. A. Verbenaceae B. *Lantana camara*
     C. Lantana D. Pakakrong
     E. Tropical America F. Rayong
     G. An aromatic, sprawling shrub with orange and yellow flower heads, it has prickly stems, serrate leaves and clusters of black, berry-like fruits. The orange-flowered variety has naturalized in waste places and scrub. Several hybrid varieties are popularly grown in gardens.

475. A. Verbenaceae B. *Lippia nodiflora*
     C. Cape Weed E. Tropical America
     F. Thonburi
     G. A creeping, branched herb which roots at the nodes. The leaves are opposite, obovate, wedge-shaped at the base and sharply-toothed. The flowers occur in axillary, cylindric, compact heads and are light pink or white and very small.

476. A. Verbenaceae B. *Stachytarpheta indica*
     C. False Verbena E. Tropical America
     F. Ranong
     G. An erect herb with opposite, serrate leaves. The small, violet-blue flowers occur on long spikes.

477. A. Zygophyllaceae B. *Tribulus terrestris*
     C. Ground Bur-nut, Small Caltrops E. Pantropical
     F. Cha-am Beach, Phetchaburi
     G. This is an introduced, prostrate, creeping herb found on sandy soils. It has compound leaves and solitary yellow flowers which are often folded. The fruits are round and covered with sharp thorns.

TROPICAL FRUITS

478.   A. Anacardiaceae   B. *Anacardium occidentale*
       C. Cashew   D. Mamuang-Himapan, Mamuang-Maruhao
       E. Tropical America   F. Ranong
       G. A spreading tree with leathery, obovate leaves. The flowers are yellowish-pink and cluster in panicles at branch ends. The fruits are yellow when mature with a kidney-shaped nut at the base. The nut is poisonous until treated by heat, such as by roasting. Cashew trees are most often seen in coastal areas of S. Thailand.

479.   A. Anacardiaceae   B. *Bouea macrophylla*
       C. Remenia   D. Maprang
       E. Malaysia   F. Bangkok
       G. A medium sized tree with low, hanging branches, it has simple leaves in opposite pairs. The small, green flowers are clustered in panicles at the ends of branches. The orange fruit is oblong, thin-skinned with an orange, turpentine, sweet, fibrous pulp.

480.   A. Anacardiaceae   B. *Mangifera indica*
       C. Mango   D. Mamuang
       E. India   F. Saraburi
       G. A large shade tree with ovate-lanceolate leaves. The flowers are small, cream yellow or pink and grow on a panicle. The fruits are yellow or orange with a thin skin and fibrous pulp which has a sweet, turpentine taste. Some people are allergic to the sap.

481.   A. Annonaceae   B. *Annona reticulata*
       C. Sugar Apple, Bullock's Heart   D. Noi-noh
       E. Tropical America   F. Saraburi

478 Anacardium occidentale
479 Bouea macrophylla

480 Mangifera indica
481 Annona reticulata

482 Annona squamosa
483.1 Durio zibethinus
483.2 Durio zibethinus

484 Carica papaya
485.1 Baccaurea motleyana

485.2 Baccaurea motleyana
485.3 Baccaurea sapida
486 Phyllanthis acidus
487 Sauropus androgynus

488 Garcinia mangostana
489 Garcinia schomburgkiana

490 Lansium domesticum
491 Artocarpus altilis

492 Artocarpus heterophylla
493.1 Eugenia aquea

493.2 Eugenia aquea
493.3 Eugenia malaccensis
494 Psidium guajava

495 Punica granatum
496 Zizyphus jujuba
497.1 Citrus grandis

497.2 Citrus grandis
498 Citrus hystrix
499 Euphoria longan

500 Nephelium lappaceum
501 Nephelium litchi
502 Manilkara zapota

Euphorbiaceae

G. A medium-sized tree with dark green, ovate-lanceolate leaves. The flowers are small, green and occur singly or in groups of 2-3 close to the branches. The fruit is reddish when ripe and contains a white, sweet pulp with many black seeds which are not eaten.

482. A. Annonaceae B. *Annona squamosa*
 C. Custard Apple, Sweetsop D. Noi-nah
 E. Tropical America F. Nakhon Ratchasima
 G. A small-medium height tree with light green, alternate, ovate leaves and green flowers which grow from leaf axils. The green fruit has a deeply-reticulated surface and a consistency similar to *A. reticulata*.

483. A. Bombaceae B. *Durio zibethinus*
 C. Durian D. Thurian
 E. Malaysia F. Thonburi (1,2)
 G. A tall tree with a straight trunk and horizontal branches. The leaves are simple and drooping. The flowers (1) are borne in clusters, often on older branches, but usually only one or two develop into the large, green coarsely-spiked fruit (2). The round fruit has a thick rind which covers sections of gaseous-smelling pulp which has a custard-like consistency.

484. A. Caricaceae B. *Carica papaya*
 C. Papaya D. Malagoh
 E. Central America F. Nakhon Pathom
 G. A soft-stemmed, unbranched tree with large, deeply-lobed leaves forming an umbrella-like canopy. Male and female trees are separate. The white, male flowers hang on long stems and the white, female flowers grow on very short stems at leaf axils, bearing a large fruit with yellow-orange or red flesh surrounding a core of many, small, black seeds.

485. A. Euphorbiaceae B. *Baccaurea spp.*
 C. Rambai D. Mah-fai
 E. Malaysia and Indonesia F. Thonburi (1,2), Chiang Mai (3)
 G. *B. motleyana* (1) is a tree with thick foliage of glossy, simple leaves. The male flowers are small, pale yellowish green, without petals

and sweetly-scented. The female flowers on a separate tree are larger and produce hanging racemes of pale yellow fruit from older branches (2). A native species, *B. sapida* (3) grows in northern forests and bears an edible pink fruit which is more sour than the former species.

486. A. Euphorbiaceae B. *Phyllanthus acidus*
   C. Tahitian Gooseberry D. Mayom
   E. East Indies F. Bang-pa-in, Ayutthaya
   G. A small-medium height tree with bare, scarred branches and fans of light green leaves clustered at branch ends. The minute, reddish flowers cluster along the bare branches and produce a very sour, yellow, waxy fruit which is eaten raw or pickled.

487. A. Euphorbiaceae B. *Sauropus androgynus*
   D. Pak-wan E. Native to Thailand
   F. Ranong
   G. A spindly shrub about 2 ft. tall which is found growing next to forest streams and is cultivated as well. It has small flowers with a deep red calyx. The calyx remains attached to the ribbed fruit which is light green, turning white with a pink tinge when ripe. The fruit is reported to be edible and is eaten with *nam prik*. When tasted by the author, it appeared to have a spongy pulp and was quite tasteless.

488. A. Guttiferae B. *Garcinia mangostana*
   C. Mangosteen D. Mangkoot
   E. Malaysia F. Bangkok
   G. A shade tree of medium height with simple leaves in opposite pairs. The tree exudes a yellow sap. The flowers are fleshy and pinkish in color with four curved sepals and four petals. The fruit is round, purple and has a thick, soft, fibrous rind which covers the edible, sweet, white, fleshy, inner segments. Has been called the Queen of Fruits for its delicious flavor.

489. A. Guttiferae B. *Garcinia schomburgkiana*
   D. Madahn E. Native to Thailand
   F. Thonburi

Myrtaceae

G. A medium-height tree with leathery leaves, dark green-glossy on top and light green-dull on the underside. The small, cream-colored flowers occur along the branches and produce and elongated, glossy, green fruit which is sour. It is eaten raw or pickled. A native forest tree, it is also cultivated.

490. A. Meliaceae   B. *Lansium domesticum*
     D. Langsat   E. Java and Malaysia
     F. Bangkok
     G. A medium-height tree with large, simple, alternate leaves. The flowers have five yellow petals and occur on hanging spikes from the older branches. The fruits are a muddy yellow color and have a translucent, white pulp which is sour-sweet. The seeds are green and very bitter, and are not eaten.

491. A. Moraceae   B. *Artocarpus altilis*
     C. Breadfruit   D. Sahkae
     E. Malaysia and the Pacific   F. Bang-pa-in, Ayutthaya
     G. This is an infertile species of breadfruit which is edible, starchy and seedless. It is a large tree with deeply-lobed leaves and is often planted as an ornamental shade tree in addition to its value for its fruit.

492. A. Moraceae   B. *Artocarpus heterophylla*
     C. Jackfruit   D. Kanoon
     E. Indo-Malaysia   F. Nakhon Ratchasima
     G. A large fruit with a slight-spiny surface which contains many phalanges of perfumy, crisp flesh. The fruit grows from the trunk or older branches of the medium-tall tree.

493. A. Myrtaceae   B. *Eugenia spp.*
     C. Rose Apple, Malay Apple   D. Chompoo
     E. India and Malaysia   F. Thonburi (1,2,3)
     G. There are several species of *Eugenia* with edible fruit. They are generally medium-height trees with filament-puff flowers. *E. aquea* has cream-colored flowers (1) and red-pink, fleshy fruit (2). The larger-fruited *E. malaccensis* (3) has bright red flowers and maroon fruits.

TROPICAL FRUITS

494. A. Myrtaceae B. *Psidium guajava*
     C. Guava D. Farang, Farang Vietnam
     E. Tropical America F. Thonburi
     G. A small tree with smooth, peeling bark and opposite, rough-textured leaves. The flowers are white and attractive. The fruit is green or yellow when ripe with white or pink pulp depending upon the hybrid variety.

495. A. Punicaceae B. *Punica granatum*
     C. Pomegranate D. Tuptim
     E. Iran F. Bangkok
     G. A shrubby, small tree with short spines and small, glossy leaves. It is often planted as an ornamental for its showy, red-orange flowers. The round, yellow, orange or red fruit has many seeds, embedded in a gelatinous, sweet, edible pulp.

496. A. Rhamnaceae B. *Zizyphus jujuba*
     C. Jujube D. Phut-sah-chin
     E. Tropical Asia F. Samut Prakan
     G. A small-medium-height tree with short thorns. The leaves are alternate, elliptic with a slightly-toothed margin. Small, green and white flowers cluster at leaf axils and produce a round, greenish-yellow fruit with an apple-like pulp and a single, hard stone. Can be found cultivated or in the wild.

497. A. Rutaceae B. *Citrus grandis*
     C. Pomelo, Shaddock D. Som-oh
     E. Thailand, Indochina and Malaysia F. Chiang Mai (1,2)
     G. A small tree with ovate leaves with a small wing. The flowers (1) are white and fragrant. The fruit (2) is large, green or yellow and has a thick skin and white pith covering succulent sections of white or pink pulp.

498. A. Rutaceae B. *Citrus hystrix*
     C. Porcupine Orange D. Mahkroot
     E. Indonesia F. Nakhon Rathchasima
     G. A small-medium-height tree with thorns and distinctive leaves which appear to be divided, but are actually an ovate leaf and expanded

Sapotaceae

wing which is nearly as large as the leaf. The small flowers are white. The fruit is rough-skinned and contains little, bitter juice. Both the leaves and the rind are used as a spice.

499. A. Sapindaceae B. *Euphoria longan*
     C. Longan D. Lamyai
     E. South China F. Chiang Mai
     G. Medium-sized trees grown mostly in N. Thailand they have dark green, oblanceolate leaves, acute at both ends. The round fruits have a thin, brown skin and white, translucent pulp with a sweet, nutty flavor.

500. A. Sapindaceae B. *Nephelium lappaceum*
     C. Rambutan D. Ngork
     E. Malaysia F. Bangkok
     G. A large, bushy tree with compound leaves in which the leaflets may be alternate or opposite. The small, greenish flowers are clustered in panicles at branch ends. There are male and female flowers on separate trees. The red and yellow, oblong fruits have soft spines protruding from the skin. The pulp is firm, white, translucent, juicy and sweet. Each fruit contains one large seed.

501. A. Sapindaceae B. *Nephelium litchi*
     C. Lychee D. Linchee
     E. South China F. Bangkok
     G. A medium-height tree which bears fruit in clusters at branch ends. The fruits have a thin, warty, red skin. The pulp is white, sweet and nutty-flavored.

502. A. Sapotaceae B. *Manilkara zapota*
     C. Chicle, Sapodilla D. Lamoot
     E. Southern Mexico to Venezuela F. Phatthalung
     G. Sweet, brown fruit which grow on a small-medium-height, compact tree which is thickly-foliated with simple, dark green leaves. The flowers are small, white and grow from leaf axils.

INDEX TO SPECIES BY FAMILIES

A

Acanthaceae
  Acanthus ebracteatus   72
  Adhatoda vasica   1
  Andrographis paniculata   82
  Asystasia gangetica   87
  Asystasia intrusa   1
  Barleria cristata   1
  Barleria lupulina   2
  Barleria prionitis   42
  Barleria siamensis   1
  Clinacanthus siamensis   42
  Crossandra undulaefolia   2
  Graptophyllum pictum   2
  Justicia betonica   2
  Justicia gendarrusa   2
  Odontonema stricta   2
  Pachystachys lutea   3
  Pseuderanthemum graciflorum   42
  Psuederanthemum andersonii   3
  Psuederanthemum carruthersii   3
  Psuederanthemum setricalyx   3
  Ruellia tuberosa   87
  Sanchezia nobilis   3
  Thunbergia erecta   3
  Thunbergia fragrans   42
  Thunbergia grandiflora   4
  Thunbergia laurifolia   43
Agavaceae
  Dracaena hookeriana   4

Aizoaceae
  Sesuvium portulacastrum   72
Alismataceae
  Echinodorus cordifolius   4
Amaranthaceae
  Achyranthes aspera   87
  Alternanthera philoxeroides   82
  Alternanthera sessilis   82
  Gomphrena celosiodes   87
Amaryllidaceae
  Crinum amabile   4
  Crinum asiaticum   4
  Eucharis grandiflora   5
  Eurycles amboinensis   5
  Hymenocallis littoralis   5
  Pancratium zeylanicum   5
  Polianthes tuberosa   5
  Zephyranthes spp.   5
Anacardiaceae
  Anacardium occidentale   104
  Bouea macrophylla   104
  Mangifera indica   104
  Melanorrhoea usitata   43
Annonaceae
  Annona reticulata   104
  Annona squamosa   105
  Artabotrys siamensis   6
  Cananga odorata   6
  Desmos chinensis   6
  Rauwenhoffia siamensis   43

INDEX TO SPECIES BY FAMILIES

Uvaria macrophylla 43
Apocynaceae
  Aganosma marginata 43
  Allamanda spp. 6
  Beaumontia grandiflora 7
  Carissa carandas 7
  Catharanthus roseus 87
  Cerbera manghas 6
  Cerbera odollam 72
  Ervatamia coronaria 7
  Holarrhena densiflora 7
  Nerium spp. 7
  Odontadenia speciosa 8
  Plumeria spp. 8
  Strophanthus gratus 8
  Thevetia peruviana 8
  Vallaris glabra 8
  Wrightea dubia 44
  Wrightia religiosa 9
Araceae
  Amorphophallus sp. 44
  Lasia spinosa 44
  Pistia stratiotes 82
  Scindapsus siamense 44
  Spathiphyllum clevelandii 9
  Typhonium trilobatum 88
Araliaceae
  Aralia armata 44
Aristolochiaceae
  Apama tomentosa 45
  Aristolochia galeata 9
Asclepiadaceae
  Calotropis gigantea 88
  Oxystelma esculentum 88
  Telosma minor 45
  Tylophora indica 72

B
Balanophoraceae
  Balanophora latisepala 45
Balsaminaceae
  Impatiens masonii 66
Begoniaceae
  Begonia.spp. 45
Bignoniaceae
  Adenium obesum 9
  Crescentia alata 9
  Crescentia cujete 10
  Jacaranda mimosaefolia 10
  Millingtonia hortensis 10
  Pseudocalymma alliaceum 10
  Spathodea campanulata 10
  Tabebuia spp. 11
  Tecoma stans 11
Bixaceae
  Bixa orellana 11
Bombaceae
  Durio zibethinus 105
Boraginaceae
  Cordia dentata 11
  Cordia sebestena 12
  Cordia subcordata 73
  Heliotropium indicum 88
Bromeliaceae
  Pitcairnea flammea 12
Butomaceae
  Hydrocleys nymphoides 12
  Limnocharis flava 82

C
Cactaceae
  Opuntia vulgaris 73
  Pereskia spp. 12

111

INDEX TO SPECIES BY FAMILIES

Cannaceae
    Canna generalis   12
Capparidaceae
    Capparis micracantha   13
    Cleome speciosa   13
    Cleome viscosa   89
    Crataeva spp.   13
    Gynandropsis pentaphylla   89
Caprifoliaceae
    Lonicera japonica   13
    Sambucus canadensis   14
Caricaceae
    Carica papaya   105
Casuarinaceae
    Casuarina equisetifolia   73
Chenopodiaceae
    Suaeda maritima   73
Chloranthaceae
    Chloranthus inconspicuus   46
Cochlospermaceae
    Cochlospermum religiosum   14
Combretaceae
    Calycopteris floribunda   46
    Lumnitzera racemosa   73
    Quisqualis indica   14
    Terminalia catappa   14
Commelinaceae
    Commelina spp.   83
    Forrestia griffithii   46
    Pollia thyrsifolia   46
Compositae
    Eclipta prostrata   89
    Elephantopus scaber   89
    Erechthites valerianifolia   89
    Eupatorium odoratum   89
    Pluchea indica   74
    Spilanthes acmella   90
    Tridax procumbens   90
    Vernonia cinerea   90
    Vernonia sp.   66
    Wedelia biflora   74
    Wedelia prostrata   46
Connaraceae
    Cnestis palala   47
Convolvulaceae
    Argyreia mollis   47
    Argyreia nervosa   14
    Argyreia splendens   66
    Calonyction album   74
    Cuscuta chinensis   90
    Hewittia sublobata   90
    Ipomoea aquatica   83
    Ipomoea carnea   15
    Ipomoea digitata   91
    Ipomoea indica   91
    Ipomoea obscura   91
    Ipomoea pes-caprae ssp.
      brasiliensis   74
    Ipomoea quamoclit   15
    Ipomoea stolonifera   74
    Merremia gemella   91
    Merremia hirta   91
Cucurbitaceae
    Coccinea indica   91
    Momordica charantia   92
    Momordica cochinchinensis   47
    Trichosanthes cucumerina   92
Cycadaceae
    Cycas pectinata   66

D
Dilleniaceae

INDEX TO SPECIES BY FAMILIES

Dillenia indica   15
Dillenia spp.   47
Dillenia suffruticosa   15
Tetracera indica   48
Tetracera loureiri   15
Dioscoreaceae
   Dioscorea hispida   48
   Dioscorea pyrifolia   48
Dipterocarpaceae
   Dipterocarpus obtusifolius   67
   Shorea roxburghii   48

E
Elaeocarpaceae
   Elaeocarpus floribundus   49
Ericaceae
   Rhododendron simsii   16
Euphorbiaceae
   Acalypha hispida   16
   Acalypha indica   92
   Acalypha wilkesiana   16
   Baccaurea spp.   105
   Breynia spp.   49
   Euphorbia cyathopora   92
   Euphorbia geniculata   93
   Glochidion littorale   75
   Homonoia riparia   49
   Hura crepitans   16
   Jatropha curcas   17
   Jatropha gossypifolia   93
   Jatropha integerrima   17
   Jatropha podagrica   17
   Phyllanthus acidus   106
   Phyllanthus pulcher   49
   Phyllanthus reticulatus   93
   Sauropus androgynus   106
   Sauropus compressus   49

   Synostemon bacciformis   75

F
Flagellariaceae
   Flagellaria indica   50

G
Gentianaceae
   Exacum tetragonum   67
   Nymphoides aurantiaca   17
   Swertia angustifolia   67
Gesneriaceae
   Aeschynanthus marmoratus   50
Goodeniaceae
   Scaevola taccada   75
Gramineae
   Coix lachryma-jobi   93
   Spinifex littoreus   75
Guttiferae
   Garcinia mangostana   106
   Garcinia schomburgkiana   106
   Messua ferrea   17
   Ochrocarpus siamensis   75

H
Hypoxidaceae
   Curculigo orchoides   50

I
Iridaceae
   Belacanda chinensis   18

L
Labiatae
   Anisomeles indica   93
   Hyptis capitata   94
   Hyptis suaveolens   94

INDEX TO SPECIES BY FAMILIES

Leonotis nepetifolia   94
Leucas aspera   94
Leucas lavendulifolia   67
Ocimum sanctum   94
Orthosiphon grandiflorus   18
Scutellaria discolor   50
Lauraceae
   Cassytha filiformis   76
Lecythidaceae
   Barringtonia racemosa   76
   Couroupita gianensis   18
Leeaceae
   Leea acuminata   67
   Leea rubra   51
Leguminosae
   Abrus precatorius   76
   Acacia auriculaeformis   18
   Acacia catechu   51
   Acacia spp.   51
   Adenanthera pavonina   51
   Aeschynomene indica   83
   Alysicarpus nummularifollus   95
   Atylosia volubilis   67
   Bauhinia bassacensis   52
   Bauhinia involucellata   52
   Bauhinia pottsii   52
   Bauhinia pulla   52
   Bauhinia spp.   18
   Bauhinia winitii   19
   Brownea grandiceps   19
   Butea monosperma   19
   Caesalpinia coriaria   19
   Caesalpinia mimosoides   52
   Caesalpinia pulcherrima   20
   Caesalpinia sappan   76
   Calliandra spp.   20
   Canavalia lineata   77

   Canavalia rosea   95
   Cassia bakeriana   20
   Cassia fistula   20
   Cassia garrettiana   52
   Cassia glauca   21
   Cassia mimosoides   68
   Cassia occidentalis   95
   Cassia spectabilis   21
   Cassia timoriensis   21
   Cassia tora   95
   Centrosema pubescens   95
   Clitoria ternatea   96
   Crotalaria chinensis   68
   Crotalaria mucronata   96
   Crotalaria sessilliflora   68
   Delonix regia   21
   Derris scandens   77
   Derris trifoliata   77
   Desmanthus virgatus   96
   Desmodium amethystinum   68
   Desmodium capitatum   96
   Desmodium pulchellum   53
   Desmodium triquetrum   96
   Dunbaria longeracemosa   68
   Entada phaseoloides   53
   Eriosema chinense   69
   Erythrina crista-galli   21
   Erythrina indica   22
   Gliricidia sepium   22
   Indigofera hirsuta   77
   Leucaena leucocephala   22
   Maniltoa gemmipara   22
   Milletia atropupurea   53
   Mimosa pigra   97
   Mimosa pudica   97
   Moghonia strobilifera   53
   Mucuna bennettii   22

INDEX TO SPECIES BY FAMILIES

Mucuna pruriens 97
Parkia speciosa 53
Parkinsonia aculeata 23
Peltophorum inerme 23
Phaseolus calcaratus 97
Phaseolus lathyroides 97
Phyllocarpus septentrionalis 23
Pterocarpus indicus 23
Pueraria phaseoloides 69
Samanea saman 23
Saraca bijuga 24
Sesbania roxburghii 83
Seshania grandiflora 24
Tamarindus indica 24
Tephrosia purpurea 77
Uraria crinita 98
Liliaceae
    Asparagus racemosus 54
    Dracaena lourieri 69
    Gloriosa superba 77
    Hemerocallis fulva 24
    Smilax sp. 54
Loganiaceae
    Buddleia madagascariense 24
Loranthaceae
    Dendrophthoe sp. 54
Lythraceae
    Lagerstroemia indica 25
    Lagerstroemia speciosa 25
    Pemphis acidula 78

M
Magnoliaceae
    Magnolia coco 25
    Michelia longifolia 25
    Talauma candollii 26
Malpighiaceae
    Galphimia glauca 26
    Hiptage benghalensis 26
    Malpighia coccigera 26
    Stigmaphyllon littorale 26
    Tristellateia australasiae 26
Malvaceae
    Abelmoschus moschatus 27, 98
    Abutilon indicum 98
    Althaea rosea 27
    Decashistia parviflora 69
    Gossypium barbadense 27
    Hibiscus moscheutos 27
    Hibiscus mutabilis 28
    Hibiscus rosa-sinensis 28
    Hibiscus schizopetalus 28
    Hibiscus syriacus 28
    Hibiscus tiliaceus 78
    Hibisus sabdariffa 98
    Malachra capitata 98
    Malvaviscus spp. 28
    Sida acuta 99
    Sida retusa 99
    Thespesia populnea 78
    Urena lobata 99
Marantaceae
    Thalia geniculata 29
Melastomaceae
    Medinilla spp. 29
    Melastoma normale 54
    Osbeckia chinensis 69
    Sonerila deflexa 55
Meliaceae
    Aglaia odorata 29
    Lansium domesticum 107
    Melia azedarach 29
Menispermaceae
    Cyclea peltata 55

INDEX TO SPECIES BY FAMILIES

Moraceae
    Artocarpus altilis   107
    Artocarpus heterophylla   107
    Broussonetia papyrifera   55
    Ficus benjamina   55
    Ficus hirta   55
    Ficus racemosa   56
    Ficus spp.   30
Musaceae
    Musa rosacea   30
Myrsinaceae
    Ardisia crispa   56
    Ardisia pilosa   70
Myrtaceae
    Callistemon lanceolatus   30
    Eucalyptis sp.   30
    Eugenia spp.   107
    Melaleuca leucadendron   84
    Psidium guajava   108
    Rhodomyrtus tomentosa   56

N
Nepenthaceae
    Nepenthes gracilis   99
Nyctaginaceae
    Boerhaavia spp.   99
    Bougainvillea spectabilis   31
Nymphaeaceae
    Nelumbo nucifera   84
    Nymphaea lotus   84
    Victoria amazonica   31

O
Ochnaceae
    Ochna integerrima   31
Olacaceae
    Olax scandens   78

Oleaceae
    Jasminum bifarium   78
    Jasminum sambac   31
    Nyctanthes arbor-tristis   31
Onagraceae
    Fuchsia x hybrida   32
    Jussiaea linifolia   84
    Ludwigia adscendens   85
    Ludwigia octovalvis   85
Orchidaceae
    Arundina graminifolia   70
    Dendrobium spp.   56
    Habenaria spp.   70
    Paphiopedilum spp.   56
    Phaius tankervilliae   57
    Spathoglottis plicata   57
Orobanchaceae
    Aeginetia indica   57
    Christisonia siamensis   70
Oxalidaceae
    Biophytum sensitivum   100
    Oxalis corniculata   100
    Oxalis rosea   32

P
Palmae
    Calamus sp.   58
    Caryota mitis   32
    Licuala grandis   32
    Licuala spinosa   58
    Nypa fruticans   79
Pandanaceae
    Pandanus odoratissimus   79
Papaveraceae
    Papaver somniferum   32
Passifloraceae
    Passiflora foetida   100

INDEX TO SPECIES BY FAMILIES

Passiflora laurifolia   33
Phytolaccaceae
   Rivinia humilis   100
Piperaceae
   Peperomia pellucida   58
Plumbaginaceae
   Plumbago auriculata   33
Polygalaceae
   Salomonia ciliata   100
Polygonaceae
   Antigonon leptopus   33
   Polygonum barbatum   58
   Polygonum pulchrum   85
Pontederiaceae
   Eichhornia crassipes   85
   Monochoria hastata   86
Portulaceae
   Portulaca pilosa   79
Punicaceae
   Punica granatum   108

R
Ranunculaceae
   Clematis spp.   33
Rhamnaceae
   Colubrina asiatica   58
   Zizyphus jujuba   108
Rosaceae
   Rosa hybrida   33
Rubiaceae
   Anthocephalus cadamba   34
   Argrostemma sp.   59
   Gardenia jasminoides   34
   Guettarda speciosa   79
   Hamelia patens   34
   Ixora spp.   34, 59
   Lasianthus oligoneurus   59

   Mitragyna rotundifolia   59
   Mussaenda sanderiana   59
   Mussaenda spp.   35
   Paederia tomentosa   101
   Pavetta indica   60
Rutaceae
   Citrus grandis   108
   Citrus hystrix   108
   Citrus microcarpa   80
   Murraya paniculata   35
   Ravenia spectabilis   35

S
Samydaceae
   Casearia grewiifolia   60
Sapindaceae
   Cardiospermum halicacabum   101
   Euphoria longan   109
   Nephelium lappaceum   109
   Nephelium litchi   109
Sapotaceae
   Manilkara zapota   109
   Mimusops elengi   35
Scrophulariaceae
   Buchnera cruciata   70
   Lindenbergia philippensis   101
   Russelia equisetiformis   35
   Striga asiatica   101
   Torenia fournieri   60
Simarubaceae
   Quassia amara   36
Solanaceae
   Brunfelsia hopeana   36
   Cestrum diurnum   36
   Cestrum nocturnum   36
   Datura metel   101
   Physalis angulata   102

INDEX TO SPECIES BY FAMILIES

Solandra nitida   36
Solanum torvum   102
Solanum trilobatum   102
Sonneratiaceae
　Sonneratia cascolaris   80
Sterculiaceae
　Byttneria aspera   60
　Helicteres hirsuta   80
　Helicteres isora   61
　Melochia corchorifolia   102
　Pterospermum diversifolium   37
　Sterculia foetida   37
　Waltheria americana   102
Strelitziaceae
　Heliconia spp.   37

T

Taccaceae
　Tacca leontopetaloides   61
Tiliaceae
　Colona auriculata   61
　Corchorus spp.   102
　Grewia paniculata   61
　Schoutenia peregrina   37
　Triumfetta pilosa   71
Turneraceae
　Turnera ulmifolia   38
Typhaceae
　Typha angustitolia   86

U

Umbelliferae
　Trachydium cambogianum   71

V

Verbenaceae
　Citharexylum spinosum   38

　Clerodendron colebrookianum   61
　Clerodendron fragrans   38
　Clerodendron inerme   80
　Clerodendron infortunatum   62
　Clerodendron paniculatum   62
　Clerodendron petasites   38
　Clerodendron quadriloculare   38
　Clerodendron serratum   71
　Clerodendron spicatum   62
　Clerodendron splendens   39
　Clerodendron ugandense   39
　Clerodendron villosum   62
　Clerodendron wallichii   62
　Congea tomentosa   63
　Duranta repens   39
　Gmelina philippensis   39
　Lantana camara   103
　Lippia nodiflora   103
　Petrea volubilis   39
　Premna integrifolia   81
　Sphenodesma pentandra   63
　Stachytarpheta indica   103
　Tectona grandis   63
　Vitex ovata   81
　Vitex trifolia   40
Vitaceae
　Cissus aristata   63
　Cissus quadrangularis   63

X

Xyridaceae
　Xyris pauciflora   86

Z

Zingiberaceae
　Alpinia purpurata   40
　Costus speciosus   64

INDEX TO SPECIES BY FAMILIES

Curcuma alismatifolia 40
Curcuma domestica 40
Curcuma spp. 64
Globba spp. 64
Hedychium coronarium 41

Kaempferia pulchra 65
Phaeomeria magnifica 41
Zingiber spectabile 65
Zygophyllaceae
    Tribulus terrestris 103

INDEX OF SCIENTIFIC NAMES

A

Abelmoschus moschatus   27, 98
Abrus precatorius   76
Abutilon indicum   98
Acacia auriculaeformis   18
Acacia catechu   51
Acacia spp.   51
Acalypha hispida   16
Acalypha indica   92
Acalypha wilkesiana   16
Acanthus ebracteatus   72
Achyranthes aspera   87
Adenanthera pavonina   51
Adenium obesum   9
Adhatoda vasica   1
Aeginetia indica   57
Aeschynanthus marmoratus   50
Aeschynomene indica   83
Aganosma marginata   43
Aglaia odorata   29
Allamanda spp.   6
Alpinia purpurata   40
Alternanthera philoxeroides   82
Alternanthera sessilis   82
Althaea rosea   27
Alysicarpus nummularifollus   95
Amorphophallus sp.   44
Anacardium occidentale   104
Andrographis paniculata   82
Anisomeles indica   93

Annona reticulata   104
Annona squamosa   105
Anthocephalus cadamba   34
Antigonon leptopus   33
Apama tomentosa   45
Aralia armata   44
Ardisia crispa   56
Ardisia pilosa   70
Argrostemma sp.   59
Argyreia mollis   47
Argyreia nervosa   14
Argyreia splendens   66
Aristolochia galeata   9
Artabotrys siamensis   6
Artocarpus altilis   107
Artocarpus heterophylla   107
Arundina graminifolia   70
Asparagus racemosus   54
Asystasia gangetica   87
Asystasia intrusa   1
Atylosia volubilis   67

B

Baccaurea spp.   105
Balanophora latisepala   45
Barleria cristata   1
Barleria lupulina   2
Barleria prionitis   42
Barleria siamensis   1
Barringtonia racemosa   76

INDEX OF SCIENTIFIC NAMES

Bauhinia bassacensis 52
Bauhinia involucellata 52
Bauhinia pottsii 52
Bauhinia pulla 52
Bauhinia spp. 18
Bauhinia winitii 19
Beaumontia grandiflora 7
Begonia.spp. 45
Belacanda chinensis 18
Biophytum sensitivum 100
Bixa orellana 11
Boerhaavia spp. 99
Bouea macrophylla 104
Bougainvillea spectabilis 31
Breynia spp. 49
Broussonetia papyrifera 55
Brownea grandiceps 19
Brunfelsia hopeana 36
Buchnera cruciata 70
Buddleia madagascariense 24
Butea monosperma 19
Byttneria aspera 60

C

Caesalpinia coriaria 19
Caesalpinia mimosoides 52
Caesalpinia pulcherrima 20
Caesalpinia sappan 76
Calamus sp. 58
Calliandra spp. 20
Callistemon lanceolatus 30
Calonyction album 74
Calotropis gigantea 88
Calycopteris floribunda 46
Cananga odorata 6
Canavalia lineata 77
Canavalia rosea 95

Canna generalis 12
Capparis micracantha 13
Cardiospermum halicacabum 101
Carica papaya 105
Carissa carandas 7
Caryota mitis 32
Casearia grewiifolia 60
Cassia bakeriana 20
Cassia fistula 20
Cassia garrettiana 52
Cassia glauca 21
Cassia mimosoides 68
Cassia occidentalis 95
Cassia spectabilis 21
Cassia timoriensis 21
Cassia tora 95
Cassytha filiformis 76
Casuarina equisetifolia 73
Catharanthus roseus 87
Centrosema pubescens 95
Cerbera manghas 6
Cerbera odollam 72
Cestrum diurnum 36
Cestrum nocturnum 36
Chloranthus inconspicuus 46
Christisonia siamensis 70
Cissus aristata 63
Cissus quadrangularis 63
Citharexylum spinosum 38
Citrus grandis 108
Citrus hystrix 108
Citrus microcarpa 80
Clematis spp. 33
Cleome speciosa 13
Cleome viscosa 89
Clerodendron colebrookianum 61
Clerodendron fragrans 38

INDEX OF SCIENTIFIC NAMES

Clerodendron inerme   80
Clerodendron infortunatum   62
Clerodendron paniculatum   62
Clerodendron petasites   38
Clerodendron quadriloculare   38
Clerodendron serratum   71
Clerodendron spicatum   62
Clerodendron splendens   39
Clerodendron ugandense   39
Clerodendron villosum   62
Clerodendron wallichii   62
Clinacanthus siamensis   42
Clitoria ternatea   96
Cnestis palala   47
Coccinea indica   91
Cochlospermum religiosum   14
Coix lachryma-jobi   93
Colona auriculata   61
Colubrina asiatica   58
Commelina spp.   83
Congea tomentosa   63
Corchorus spp.   102
Cordia dentata   11
Cordia sebestena   12
Cordia subcordata   73
Costus speciosus   64
Couroupita gianensis   18
Crataeva spp.   13
Crescentia alata   9
Crescentia cujete   10
Crinum amabile   4
Crinum asiaticum   4
Crossandra undulaefolia   2
Crotalaria chinensis   68
Crotalaria mucronata   96
Crotalaria sessilliflora   68
Curculigo orchoides   50

Curcuma alismatifolia   40
Curcuma domestica   40
Curcuma spp.   64
Cuscuta chinensis   90
Cycas pectinata   66
Cyclea peltata   55

D

Datura metel   101
Decashistia parviflora   69
Delonix regia   21
Dendrobium spp.   56
Dendrophthoe sp.   54
Derris scandens   77
Derris trifoliata   77
Desmanthus virgatus   96
Desmodium amethystinum   68
Desmodium capitatum   96
Desmodium pulchellum   53
Desmodium triquetrum   96
Desmos chinensis   6
Dillenia indica   15
Dillenia spp.   47
Dillenia suffruticosa   15
Dioscorea hispida   48
Dioscorea pyrifolia   48
Dipterocarpus obtusifolius   67
Dolichos lablab   96
Dracaena hookeriana   4
Dracaena lourieri   69
Dunbaria longeracemosa   68
Duranta repens   39
Durio zibethinus   105

E

Echinodorus cordifolius   4
Eclipta prostrata   89

INDEX OF SCIENTIFIC NAMES

Eichhornia crassipes 85
Elaeocarpus floribundus 49
Elephantopus scaber 89
Entada phaseoloides 53
Erechthites valerianifolia 89
Eriosema chinense 69
Ervatamia coronaria 7
Erythrina crista-galli 21
Erythrina indica 22
Eucalyptis sp. 30
Eucharis grandiflora 5
Eugenia spp. 107
Eupatorium odoratum 89
Euphorbia cyathopora 92
Euphorbia geniculata 93
Euphoria longan 109
Eurycles amboinensis 5
Exacum tetragonum 67

F

Ficus benjamina 55
Ficus hirta 55
Ficus racemosa 56
Ficus spp. 30
Flagellaria indica 50
Forrestia griffithii 46
Fuchsia x hybrida 32

G

Galphimia glauca 26
Garcinia mangostana 106
Garcinia schomburgkiana 106
Gardenia jasminoides 34
Gliricidia sepium 22
Globba spp. 64
Glochidion littorale 75
Gloriosa superba 77

Gmelina philippensis 39
Gomphrena celosiodes 87
Gossypium barbadense 27
Graptophyllum pictum 2
Grewia paniculata 61
Guettarda speciosa 79
Gynandropsis pentaphylla 89

H

Habenaria spp. 70
Hamelia patens 34
Hedychium coronarium 41
Heliconia spp. 37
Helicteres hirsuta 80
Helicteres isora 61
Heliotropium indicum 88
Hemerocallis fulva 24
Hewittia sublobata 90
Hibiscus moscheutos 27
Hibiscus mutabilis 28
Hibiscus rosa-sinensis 28
Hibiscus schizopetalus 28
Hibiscus syriacus 28
Hibiscus tiliaceus 78
Hibisus sabdariffa 98
Hiptage benghalensis 26
Holarrhena densiflora 7
Homonoia riparia 49
Hura crepitans 16
Hydrocleys nymphoides 12
Hymenocallis littoralis 5
Hyptis capitata 94
Hyptis suaveolens 94

I

Impatiens masonii 66
Indigofera hirsuta 77

123

INDEX OF SCIENTIFIC NAMES

Ipomoea aquatica   83
Ipomoea carnea   15
Ipomoea digitata   91
Ipomoea indica   91
Ipomoea obscura   91
Ipomoea pes-caprae ssp.
   brasiliensis   74
Ipomoea quamoclit   15
Ipomoea stolonifera   74
Ixora spp.   34, 59

J
Jacaranda mimosaefolia   10
Jasminum bifarium   78
Jasminum sambac   31
Jatropha curcas   17
Jatropha gossypifolia   93
Jatropha integerrima   17
Jatropha podagrica   17
Jussiaea linifolia   84
Justicia betonica   2
Justicia gendarrusa   2

K
Kaempferia pulchra   65

L
Lagerstroemia indica   25
Lagerstroemia speciosa   25
Lansium domesticum   107
Lantana camara   103
Lasia spinosa   44
Lasianthus oligoneurus   59
Leea acuminata   67
Leea rubra   51
Leonotis nepetifolia   94
Leucaena leucocephala   22

Leucas aspera   94
Leucas lavendulifolia   67
Licuala grandis   32
Licuala spinosa   58
Limnocharis flava   82
Lindenbergia philippensis   101
Lippia nodiflora   103
Lonicera japonica   13
Ludwigia adscendens   85
Ludwigia octovalvis   85
Lumnitzera racemosa   73

M
Magnolia coco   25
Malachra capitata   98
Malpighia coccigera   26
Malvaviscus spp.   28
Mangifera indica   104
Manilkara zapota   109
Maniltoa gemmipara   22
Medinilla spp.   29
Melaleuca leucadendron   84
Melanorrhoea usitata   43
Melastoma normale   54
Melia azedarach   29
Melochia corchorifolia   102
Merremia gemella   91
Merremia hirta   91
Messua ferrea   17
Michelia longifolia   25
Milletia atropupurea   53
Millingtonia hortensis   10
Mimosa pigra   97
Mimosa pudica   97
Mimusops elengi   35
Mitragyna rotundifolia   59
Moghonia strobilifera   53

INDEX OF SCIENTIFIC NAMES

Momordica charantia 92
Momordica cochinchinensis 47
Monochoria hastata 86
Mucuna bennettii 22
Mucuna pruriens 97
Murraya paniculata 35
Musa rosacea 30
Mussaenda sanderiana 59
Mussaenda spp. 35

N

Nelumbo nucifera 84
Nepenthes gracilis 99
Nephelium lappaceum 109
Nephelium litchi 109
Nerium spp. 7
Nyctanthes arbor-tristis 31
Nymphaea lotus 84
Nymphoides aurantiaca 17
Nypa fruticans 79

O

Ochna integerrima 31
Ochrocarpus siamensis 75
Ocimum sanctum 94
Odontadenia speciosa 8
Odontonema stricta 2
Olax scandens 78
Opuntia vulgaris 73
Orthosiphon grandiflorus 18
Osbeckia chinensis 69
Oxalis corniculata 100
Oxalis rosea 32
Oxystelma esculentum 88

P

Pachystachys lutea 3
Paederia tomentosa 101
Pancratium zeylanicum 5
Pandanus odoratissimus 79
Papaver somniferum 32
Paphiopedilum spp. 56
Parkia speciosa 53
Parkinsonia aculeata 23
Passiflora foetida 100
Passiflora laurifolia 33
Pavetta indica 60
Peltophorum inerme 23
Pemphis acidula 78
Peperomia pellucida 58
Pereskia spp. 12
Petrea volubilis 39
Phaeomeria magnifica 41
Phaius tankervilliae 57
Phaseolus calcaratus 97
Phaseolus lathyroides 97
Phyllanthus acidus 106
Phyllanthus pulcher 49
Phyllanthus reticulatus 93
Phyllocarpus septentrionalis 23
Physalis angulata 102
Pistia stratiotes 82
Pitcairnea flammea 12
Pluchea indica 74
Plumbago auriculata 33
Plumeria spp. 8
Polianthes tuberosa 5
Pollia thyrsifolia 46
Polygonum barbatum 58
Polygonum pulchrum 85
Portulaca pilosa 79
Premna integrifolia 81
Pseuderanthemum graciflorum 42
Pseudocalymma alliaceum 10

INDEX OF SCIENTIFIC NAMES

Psidium guajava 108
Psuederanthemum andersonii 3
Psuederanthemum carruthersii 3
Psuederanthemum setricalyx 3
Pterocarpus indicus 23
Pterospermum diversifolium 37
Pueraria phaseoloides 69
Punica granatum 108

Q
Quassia amara 36
Quisqualis indica 14

R
Rauwenhoffia siamensis 43
Ravenia spectabilis 35
Rhododendron simsii 16
Rhodomyrtus tomentosa 56
Rivinia humilis 100
Rosa hybrida 33
Ruellia tuberosa 87
Russelia equisetiformis 35

S
Salomonia ciliata 100
Samanea saman 23
Sambucus canadensis 14
Sanchezia nobilis 3
Saraca bijuga 24
Sauropus androgynus 106
Sauropus compressus 49
Scaevola taccada 75
Schoutenia peregrina 37
Scindapsus siamense 44
Scutellaria discolor 50
Sesbania roxburghii 83
Seshania grandiflora 24

Sesuvium portulacastrum 72
Shorea roxburghii 48
Sida acuta 99
Sida retusa 99
Smilax sp. 54
Solandra nitida 36
Solanum torvum 102
Solanum trilobatum 102
Sonerila deflexa 55
Sonneratia cascolaris 80
Spathiphyllum clevelandii 9
Spathodea campanulata 10
Spathoglottis plicata 57
Sphenodesma pentandra 63
Spilanthes acmella 90
Spinifex littoreus 75
Stachytarpheta indica 103
Sterculia foetida 37
Stigmaphyllon littorale 26
Striga asiatica 101
Strophanthus gratus 8
Suaeda maritima 73
Swertia angustifolia 67
Synostemon bacciformis 75

T
Tabebuia spp. 11
Tacca leontopetaloides 61
Talauma candollii 26
Tamarindus indica 24
Tecoma stans 11
Tectona grandis 63
Telosma minor 45
Tephrosia purpurea 77
Terminalia catappa 14
Tetracera indica 48
Tetracera loureiri 15

INDEX OF SCIENTIFIC NAMES

Thalia geniculata 29
Thespesia populnea 78
Thevetia peruviana 8
Thunbergia erecta 3
Thunbergia fragrans 42
Thunbergia grandiflora 4
Thunbergia laurifolia 43
Torenia fournieri 60
Trachydium cambogianum 71
Tribulus terrestris 103
Trichosanthes cucumerina 92
Tridax procumbens 90
Tristellateia australasiae 26
Triumfetta pilosa 71
Turnera ulmifolia 38
Tylophora indica 72
Typha angustitolia 86
Typhonium trilobatum 88

U
Uraria crinita 98
Urena lobata 99
Uvaria macrophylla 43

V
Vallaris glabra 8
Vernonia cinerea 90
Vernonia sp. 66
Victoria amazonica 31
Vitex ovata 81
Vitex trifolia 40

W
Waltheria americana 102
Wedelia biflora 74
Wedelia prostrata 46
Wrightea dubia 44
Wrightia religiosa 9

X
Xyris pauciflora 86

Z
Zephyranthes spp 5
Zingiber spectabile 65
Zizyphus jujuba 108

127

INDEX OF COMMON NAMES

A

African Tulip 10
Alligator Weed 82
Amazon Lily 5
American Cassia 21
Angel's Trumpet 101
Ashanti Blood 35
Asoke-khao 22
Australian Pine 73
Australian Wattle 18

B

Baloon Vine 101
Balsam Pear 92
Barbary Fig 73
Bastard Mustard 89
Beach Morning Glory 74
Beach Sunflower 74
Beaumontia 7
Bee Plant 13
Bengal Clockvine 4
Bitter Cucumber 92
Bitter Melon 92
Bitterwood 36
Black Varnish Tree 43
Blackberry Lily 18
Bleeding Heart 39
Blue Butterfly 39
Blue Hibiscus 28
Blue Morning Glory 91

Blue Pea 96
Bo Tree 30
Bougainvillea 31
Bracteate Desmodium 53
Brazilian Golden Vine 26
Bread Flower 8
Breadfruit 107
Broomrape 57
Buffalo Clover 95
Bullock's Heart 104
Bur-fruited Urena 99
Bush Teabush 94
Bush Thunbergia 3
Buttercup Tree 14
Butterfly Pea 96
Buttonweed 94

C

Calabash 9, 10
Canna Lily 12
Cannonball Tree 18
Cape Gooseberry 102
Cape of Good Hope 4
Cape Weed 103
Caper Tree 13
Caricature Plant 2
Cashew 104
Catechu Tree 51
Cat's Tail 98
Cat's Whiskers 18

INDEX OF COMMON NAMES

Cattail   86
Chaff Flower   87
Chain-of-Love   33
Champac   25
Changeable Rose   28
Chenille Plant   16
Chicle   109
China Inkberry   36
Chinaberry   29
Chinese Desmos   6
Chinese Rhododendron   16
Chinese Rice Flower   29
Clematis   33
Climbing Oleander   8
Coat Buttons   90
Cockscomb   21
Coffee Senna   95
Common Acalypha   92
Common Salomonia   100
Common Spurge   93
Copper Pod   23
Copperleaf   16
Coral Bean   76
Coral Berry   100
Coral Hibiscus   28
Coral Tree   22
Cork Tree   80
Country Mallow   98
Crab's Eye   76
Creeping Water Primrose   85
Crepe Jasmine   7
Crepe Myrtle   25
Crinum   4
Crown Flower   88
Cup-of-Gold   36
Custard Apple   105
Cycad   66

D
Dark-leaved Crossandra   2
Day Lily   24
Dayflower   83
Divi-divi   19
Dodder   90
Durian   105
Dwarf Poinciana   20
Dwarf Poinsettia   92

E
Ear-stud Flower   90
Elder   14
Elephant Apple   15
Elephant Creeper   14
Elephant's Foot   89
Elephant's Foot Yam   44

F
False Daisy   89
False Rattan   50
False Verbena   103
False Violet   50
Fan Palm   32, 58
Ficus   30
Fire Bush   34
Fire-cracker Flower   35
Fireweed   89
Fishtail Palm   32
Flame Tree   21
Flame-of-the-Forest   19
Fragrant Clerodendron   38
Frangipani   8
Fuchsia   32

G
Gardenia   34

129

INDEX OF COMMON NAMES

Garlic Vine   10
Giant Indian Milkweed   88
Giant Sensitive Plant   97
Globe Amaranth   87
Gloriosa Lily   77
Golden Chalice   36
Golden Dewdrop   39
Golden Eranthemum   2
Golden Rod   26
Golden Shower   20
Golden Water Snowflake   17
Golden-eyed Grass   50
Gout Plant   17
Greater Seabean   95
Griffith's Forrestia   46
Ground Bur-nut   103
Guava   108
Gum Tree   30

H
Hairy Triumfetta   71
Half-flower   75
Hanging Heliconia   37
Hastate-leaved Pondweed   86
Hedge-row   48
Heliconia   37
Hibiscus   28
Hibiscus Tree   78
Hollyhock   27
Honeysuckle   13
Hop-headed Barleria   2
Horse Tamarind   22
Hyacinth Bean   96

I
Indian Asystasia   87
Indian Cork Tree   10

Indian Lilac   29
Indian Marsh Fleabane   74
Indian Rose Chestnut   17
Indian Turnsole   88
Indonesian Clerodendron   62
Ironwood   73
Ivory Plant   88
Ixora   34, 59

J
Jacaranda   10
Jack-in-the-bush   89
Jackfruit   107
Japanese Canna   37
Jasmine   31, 78
Jasmine Tree   7
Jerusalem Thorn   23
Jewel Vine   77
Job's Tears   93
Joseph's Coat   16
Jujube   108

K
Kathe of India   51
Knotweed   58

L
Lady-of-the-Night   36
Lady's Slipper   56
Lantana   103
Leadwort   33
Leopard Flower   18
Lilac Eranthemum   42
Limestone Cassia   21
Lipstick Plant   11
Little Ironweed   90
Lobster Claw   37

INDEX OF COMMON NAMES

Lollipop Plant 3
Longan 109
Lotus Banana 30
Lotus Ginger 40
Love-in-a-mist 100
Lychee 109

M
Madre de Cacao 22
Magnolia 25, 26
Malay Apple 107
Malayan Dillenia 15
Malayan Groundsel 89
Mallow Rose 27
Mango 104
Mangosteen 106
Many-flowered Pollia 46
Medinilla 29
Mickey Mouse Tree 31
Milla 40
Mimosa-leaved Cassia 68
Monkey-flower Tree 23
Monkeypod 23
Monkey's Dinner Bell 16
Morning Glory 15
Musk Lime 80
Musk Mallow 27, 98
Mussaenda 35

N
Narrow-leaved Willow Herb 84
Natal Plum 7
Nipa 79

O
Oleander 7
Opium Poppy 32

Orange Jasmine 35
Orchid Tree 18
Orchid Vine 26
Organ Mountain 12

P
Pagoda Flower 62
Pandanus 79
Papaya 105
Paper Bark 84
Paper Flower 63
Paper Mulberry 55
Para Cress 90
Passionflower 33
Peepul Tree 30
Periwinkle 87
Persian Lilac 29
Philippine Ground Orchid 57
Philippine Violet 1
Physic Nut 17
Pigeonberry 39
Pink Bignonia 9
Pink Ravenia 35
Pink Shower Tree 20
Pink Tecoma 11
Pink Wood Sorrel 32
Pink-eyed Cerbera 6
Pitcher Plant 99
Plumeria 8
Polynesian Arrowroot 61
Pomegranate 108
Pomelo 108
Popping Pod 87
Porcupine Orange 108
Prayerbead 76
Prickly Pear Cactus 73
Pride of Barbados 20

INDEX OF COMMON NAMES

Pride of India   25
Purple Psuederanthemum   3
Purple Wreath   39

Q
Queen Sirikit Mussaenda   35
Quezonla   38

R
Railway Creeper   91
Rain Lily   5
Raintree   23
Rambai   105
Rambutan   109
Rangoon Creeper   14
Rataan Palm   58
Rattlebox   96
Red Crinum   4
Red Ginger   40
Red Jade Vine   22
Red Powder-puff   20
Red Sorrel   98
Red-veined Begonia   45
Redhead Powder-puff   20
Remenia   104
Rooster-flower   9
Rose   33
Rose Apple   107
Rose Myrtle   56
Rose of India   25
Rose-flowered Jatropha   17
Roselle   98
Royal Poinciana   21
Royal Water Lily   31

S
Sacred Basil   94

Sacred Lotus   84
Sage Rose   38
Sandalwood Tree   51
Sandbox Tree   16
Sandpaper Vine   39
Sapodilla   109
Sappan Wood   76
Scarlet Bean   97
Scarlet Bush   34
Scarlet Flame Bean   19
Scrambled Eggs   21
Screwpine   79
Sea Blite   73
Sea Holly   72
Sea Island Cotton   27
Sea Purslane   72
Sensitive Plant   97
Shaddock   108
Shower-of-Orchids   63
Showy Bottlebrush   30
Siamese Barleria   1
Silver Morning Glory   14
Silverbush   58
Singapore Holly   26
Sleeping Hibiscus   28
Small Caltrops   103
Sola Plant   83
Spider Flower   13, 89
Spider Lily   5
Spiderwort   83
Spike-fruited Crow's Cucumber   47
Spinifex Grass   75
Spreading Hogweed   99
Star Grass   50
Star Ipomoea   15
Straits Rhododendron   54
Sugar Apple   104

INDEX OF COMMON NAMES

Swamp Rose Mallow 27
Sweetsop 105

T
Tahitian Gooseberry 106
Tamarind 24
Teak 63
Thai Bauhinia 52
Thai Begonia 45
Thai Caper 13
Thai Ginger 65
Three-lobed Typhonium 88
Thunbergia 43
Tidal Marsh Glochidion 75
Tiger's Claw 22
Today and Tomorrow 36
Tonkin Jasmine 45
Torch Ginger 41
Tree of Sadness 31
Tropical Almond 14
Tuba Root 77
Tuberose 5
Tulip Tree 10
Turk's Cap 28
Turmeric 40

U
Upright Burhead 4

V
Village Ardisia 56
Violet Allamanda 6

W
Water Canna 29
Water Hyacinth 85
Water Lettuce 82
Water Lily 84
Water Morning Glory 83
Water Poppy 12
Water Primrose 85
Watersmart Weed 85
White Costus 64
White Flag 9
White Ginger 41
White Shrimp Plant 2
Wild Asparagus Fern 54
Wild Clary 88
Wild Eggplant 102
Wild Ginger 64
Wild Yam 48
William's Allamanda 6
Wishbone Flower 60
Witchweed 101
Wood Sorrel 100

Y
Yellow Allamanda 6
Yellow Burhead 82
Yellow Cleome 89
Yellow Elder 11
Yellow Morning Glory 91
Yellow Oleander 8
Yellow Poinciana 23
Yellow Tabebuia 11
Yesterday 36

INDEX OF THAI PLANT NAMES

A
Ah-luang   54
Ah-noi   69
Aki-tawan   71
Amae-son   4
Angab   1, 42
Asoke-khao   22
Asoke-nam   24
Asoke-sapun   19

B
Bahtavia   17
Bai-la-baht   14
Bai-nahk   3
Bai-ngun   2
Ban Phe   77
Ban-buree   6
Ban-buree-saet   8
Ban-chao   38
Ban-mai-ru-roi-pah   87
Ban-toen   8
Beep   10
Begonia   45
Boon-nahk   17
Boong   83
Bu-du Bu-lang   45
Bua   84
Bua-bah   17
Bua-farang   5
Bua-kradong   31

Bua-ngun   5
Bua-sai   84
Buap-khom   92
Buk   44
Buknga Bali   38

C
Cha Bahtavia   26
Chaba   28
Chaiya-pruek   20
Chamaliang-ban   67
Champa-tet   37
Chamuuk-plah-lai-dong   88
Chan-daeng   69
Chat-fah   62
Chat-phra-in   67, 94
Chaw-muang   39
Ching-chai   39
Ching-chee   13
Chingcho-lek   90
Chomanat   8
Chompoo   107
Chompoo-puntip   11
Chong-ko   18, 52
Chong-ko Taksin   52
Chong-nang   3
Chuan-chom   9

D
Dao-a-dung   77

INDEX OF THAI PLANT NAMES

Dharma-raksa 37
Din-saw 40
Dok-din 70
Don-ya 35
Don-ya Sirikit 35
Dtao-chang-goh 32
Dteen-ped-farang 9
Dteen-ped-nam 6, 72
Dteen-tukkae 90
Dtree-chawa 2
Duei 93
Dunyoeng 19

F
Fai-kum 14
Fai-phee 98
Fai-tet 27
Fak-khao 47
Farang 108
Farang Vietnam 108
Fin 32
Fin-nam 12
Foi-fah 46
Foi-mai 90
Fueng-fah 31

G
Gaeo 35
Gah-lah 41
Gai-fah 9
Gai-phra-law 10
Galapah-pruek 20
Gatiem-tao 10
Glahm-raed 44
Gloei 48
Gnak-plah-maw 72
Gnoen-gai 47

Grabong-pet 73
Gulap 33
Gulap-doi 16
Gulap-maulum-lueng 12
Gwuk-mae-chan 9

H
Haew-pradu 69
Hahng-kra-rok 98
Hahng-kra-rok-daeng 16
Hahng-nok-yoong 21
Hanuman-nung-taen 17
Hi-ran-yiga 7
Hieng 67
Hollyhock 27
Hu-kwang 14
Hu-ling 16
Hua-chai-naep 44

I
Intanin-nam 25

J
Jahk 79
Jahmjuree 23
Jauk 82
Jik-suan 76
Jum-kong 56
Jum-pee 25

K
Ka-thok-rok 100
Kah-dung-bai 51
Kah-fak 54
Kah-gai-daeng 2
Kah-ling 64
Kah-min-khok 40

135

INDEX OF THAI PLANT NAMES

Kahk-mahk   45
Kahng-plah   49
Kajorn   45
Kam-daeng   63
Kameng   89
Kamlang-changsan   31
Kan-soeng   58
Kanika   31
Kanoon   107
Karah   94
Kasae   53
Kasalong   38
Kathin-narong   18
Katum-nam   34
Keh-sonla   38
Khae-farang   22
Khae-khao   24
Khao-tok-taek   46
Khee-daeng   53
Khee-khrok   99
Khee-lek American   21
Khee-lek-ban   21
Khee-lek-lued   21
Khee-lek-phee   95
Khem   34, 59
Khem-khao   59, 60
Khem-muang   3
Khem-sethi   34
Khing-daeng   40
Khok-kra-om   101
Klet-plachon   53
Kluay-bua   30
Kluay-mai-nam   29
Koon   20
Kra-jio   40
Krachiap-daeng   98
Kradanga   6

Kradanga-tao   6
Krang   68
Kraproh   58
Krathin   22
Krathin-tet   51
Kratok-rok-farang   33
Kratoom-nern   59
Krua-khao-luang   66
Krua-phu-ngoen   47
Kruay   60
Kum-bok   13
Kum-nam   13
Kum-saet   11

L
Lamoot   109
Lampong   101
Lampoo   80
Lamyai   109
Lan-tom   8
Langsat   107
Leb-mue-nang   14
Lian   29
Lin-kuram   59
Linchee   109
Liu-dok   30
Luang India   11
Luang-kiriboon   3

M
Ma-hoi   92
Ma-tad   15
Madah   106
Maduea-din   43
Maduea-khon   55
Maduea-uthumpon   56
Maeng-luk   94

INDEX OF THAI PLANT NAMES

Mah-fai 105
Mah-kome 61
Mah-kua-puang 102
Maha-hongse 41
Mahing-peh 68
Mahkroot 108
Mai-thao-rusee 61
Maiyarap 97
Makam 24
Makam-bia 68
Maklam-tah-chang 51
Malagoh 105
Mali 31
Mamuang 104
Mamuang-Himapan 104
Mamuang-Maruhao 104
Mamui 97
Mangkoot 106
Mangkorn-angab-daeng 42
Maprang 104
Maprao-dao-luang 66
Mayom 106
Mok 9
Mok-daeng 44
Montha 26
Muchalin 35
Muen-doi 49

N
Naht-wua 74
Nam-daeng 7
Nam-dtao-ton 10
Nam-ki-ret 51
Nam-neh-khao 42
Nam-tao 99
Nang-yaem 38
Nang-yaem-pah 62

Ngork 109
No-rah 26
Noi-nah 105
Noi-noh 104
Nok-yoong Thai 20
Nome-chang 43
Nome-maeo 43
Non-see 23

O
Oen-galieng 71
Oon-farang 14
Orapin 19

P
Pahm-jeep 32
Pai-nam 85
Pak-boong-farang 15
Pak-boong-rua 91
Pak-boong-talay 74
Pak-kra-sank 58
Pak-krat-talay 74
Pak-nam 44
Pak-pai-nam 58
Pak-ped-nam 82
Pak-prap 83
Pak-pu-yah 52
Pak-sian 89
Pak-sian-farang 13
Pak-sian-pee 89
Pak-top Thai 86
Pak-top-chawa 85
Pak-wan 106
Pakakrong 103
Pakat-hua-waen 90
Pang-poey-nam 85
Paw-bit 61

137

INDEX OF THAI PLANT NAMES

Paw-talay 78
Payap-mohk 33
Payohm 48
Payup-maek 18
Phaeng-phuai-farang 87
Phee-sua 39
Phratat Filipine 34
Phratat-chin 35
Phratat-yai 36
Phurahong 28
Phut-jeep 7
Phut-sah-chin 108
Phut-sawn 34
Phut-toong 7
Phutta-raksa 12
Phuttachat-sam-see 36
Phuttan 28
Phuttan-rah 27
Pid-taw 49
Pikul 35
Pin-daeng 62
Ping-khao 61
Plub-pleung 4, 5
Plub-pleung-daeng 4
Po 30
Po-farang 16
Po-pahn 61
Poh-krasah 55
Poo-chompoo 20
Poo-daeng 20
Pradu 23
Pradu-daeng 23
Prayong 29
Prik-farang 100
Pro-pah 65
Puang-chompoo 33
Puang-gaeo-daeng 39
Puang-gomaen 22
Puang-kaeo-kudan 33
Puang-kram 39
Puang-muang 39
Puang-pradit 63
Puang-them-hu 70
Puang-thong-krua 26
Puang-thong-ton 26

R

Ra-ngap 97
Ra-ruen 46
Rachavadee 24
Rahng-juet 43
Rak 43
Rampon 12
Ratamah 23
Ratree 36
Raya-khao 62
Rayong 77
Rongtao-naree 56
Rongtao-naree-muang-kan 56
Rongtao-naree-rueng-prachin 56
Rot-sukon 15
Ruang-pueng 37
Ruk 88
Rumpoey 8

S

Sa-lah 18
Sa-meh-san 52
Sa-no 83
Sa-uek 91
Saba 53
Saboo-daeng 93
Saboo-dum 17
Saeng-chan-lek 2

INDEX OF THAI PLANT NAMES

Sahk   63
Sahkae   107
Sahn   15
Sahn-chang   47
Sahn-goranee   2
Sahn-hing   47
Sai   30
Sai-bai-sam-liam   30
Sai-inthanin   4
Sai-nam-pueng   13
Sai-yoi-bai-laem   55
Salaeng-pan   52
Salaeng-pan-tao   52
Salate-pahng-porn   2
Sam-roi-taw   63
Sam-sip   54
Sanied   1
Sao-sanom   55
Sarapee   75
Sataw   54
Sawn-klin   5
Sawn-klin-daeng   12
Sayoot   6
See-siad   51
Singto-dum   61
Soi-raya   29
Som-chaba   27
Som-oh   108
Sone-gahng-plah   15
Srakae-bai-dum   63
Sri-trang   10
Sumrong   37
Supanika   14
Suwana-pruek   11

T

Ta-ga-poh   67

Tabaek   25
Takrai-nam   49
Talapat Rusee   82
Tam-lueng   91
Tamyae Maeo   92
Tao-lang-lai   42
Tao-orakhon   48
Tao-thong-thuean   60
Tao-wan-priang   77
Tao-wun-yung   54
Tawp-tep-nam   77
Thong-gwao   19
Thong-lang   22
Thong-lang Hongkong   21
Thong-pan-doon   69
Thong-samrit   3
Thong-urai   11
Thurian   105
Tian-nam   85
Tien-yod   39
Tu-pruet   56
Tua-kla   95
Tua-lai   95
Tua-paep   96
Tua-phee   67, 97
Tua-sian-pah   69
Tuay-thong   36
Tuptim   108

U

Uang-atakrit   56
Uang-chan   96
Uang-chang-nao   56
Uang-din   57, 64
Uang-gnoen-gai   50
Uang-kankoke   56
Uang-prao   57

INDEX OF THAI PLANT NAMES

Uang-pueng   56
Uang-sai-prasat   56
Uang-thong   3
Utaphit   88

W
Waeo-mayurah   60
Wai   58
Wai-ling   50
Wan Mahamehk   64
Wan-hahng-chang   18
Wan-mahachoke   5
Wan-nang-lom   5
Wan-prai-dum   65
Wan-prao   50
Wan-torani-san   49
Wasana   4

Y
Yah-dok-khao   89
Yah-fai-nok-khum   89
Yah-farangset   89
Yah-hoon-hai   49
Yah-hua-toh   94
Yah-khao-gum   57
Yah-nguang-chang   88
Yah-nuat-suea   62
Yah-sam-wan   90
Yah-toop   86
Yah-yah   1
Yee-hoob   25
Yee-toh   7
Yee-toh Penang   70
Yotakah   18

BIBLIOGRAPHY

Henderson, M.R. *Malayan Wild Flowers*, The Malayan Nature Society, Singapore, 1979.

Merrill, E.D. *A Flora of Manila*, Bureau of Science, Manila, 1912.

Merrill, E.D. *Plant Life of the Pacific World*, The Macmillan Co., 1945.

Moore, Philip H. and Krizman, Richard D. *Field and Garden Plants of Guam*, University of Guam, 1981.

Moore, Philip H. and McMakin, Patrick D. *Plants of Guam*, University of Guam, 1979.

Pancho, Juan V. and Soerjani, Mohamad. *Aquatic Weeds of Southeast Asia*, University of the Philippines at Los Banos College, Laguna and the SEAMEO Regional Center for Tropical Biology, Bogor, Indonesia, 1978.

Pinratana, Bro. Amnuay. *Flowers in Thailand*, Vols. 1-11, Viratham Press, Bangkok, 1973-84.

Pongpangan, Somchit and Poobrasert, Suparb. *Edible and Poisonous Plants in Thai Forests*, Science Society of Thailand, 1971.

Ridley, Henry N. *The Flora of the Malay Peninsula*, L. Reeve & Co., Ltd., London, 1922.

Smitinand, Tem. *Thai Plant Names*, Royal Forest Department, Bangkok, 1980.

Stone, Benjamin C. "The Flora of Guam", *Micronesica*, Vol. 6, University of Guam, 1970.

Suvatti, Chote. *Flora of Thailand*, Royal Institute, Bangkok, 1978.